ROCKET BOYS

ROCKET BOYS

A Memoir

Homer H. Hickam, Jr.

DELACORTE PRESS

3/05/99 *AGG-4377*

Y
B
HICKAM, H. /
HIC

Published by
Delacorte Press
Bantam Doubleday Dell Publishing Group, Inc.
1540 Broadway
New York, New York 10036

It's All in the Game by Charles Gates Dawes and Carl Sigman. Lyrics reprinted courtesy of Major Songs (ASCAP) c/o The Songwriters Guild of America © 1951, and Warner Bros. Publications, Inc. Rights for the British Reversionary Territories controlled by Memory Lane Music Limited, London. All Rights Reserved. Used by Permission.

Love Is a Many Splendored Thing by Paul Francis Webster and Sammy Fain. © 1955 Twentieth Century Music Corporation. © Renewed and Assigned to EMI Miller Catalog, Inc. All Rights Reserved. Used by Permission of Warner Bros. Publications, Inc.

The author has sought to trace ownership and, when necessary, obtain permission for quotations included in this book. Occasionally he has been unable to determine or locate the author of a quote. In such instances, if an author of a quotation wishes to contact the author of this book, he or she should contact him through the publisher.

Library of Congress Cataloging in Publication Data

Hickam, Homer H., 1943–

Rocket boys

p.   cm.

ISBN 0-385-33320-X

1. Hickam, Homer H., 1943–   . 2. Aerospace engineers—United States—Biography.

TL789.85.H53A3   1998

629.1′092′273—dc21

[B]   98-19304

CIP

Designed by Brian Mulligan

Manufactured in the United States of America

Published simultaneously in Canada

October 1998

10 9 8 7 6 5 4 3

BVG

To Mom and Dad
And the people of Coalwood

AUTHOR'S NOTE

THE ROCKET BOYS of the Big Creek Missile Agency and their lives and times were real, but it should be mentioned that I have used a certain author's license in telling their story. While I have used the actual names for each of the boys and my parents and most of the people in this book, I have used pseudonyms for others and also sometimes combined two or more people into one when I felt it necessary for clarification and simplification. I have also taken certain liberties in the telling of the story, particularly having to do with the precise sequence of events and who may have said what to whom. Nevertheless, my intention in allowing this narrative to stray from strict nonfiction was always to illuminate more brightly the truth.

ACKNOWLEDGMENTS

I OWE A DEBT OF GRATITUDE to many people for this book. First, I extend my infinite appreciation to Mickey Freiberg of The Artists Agency in Los Angeles for recognizing the value of this work from its first glimmering. It was his belief in the story, and his confidence in my ability to write it, that gave me the opportunity I needed to proceed. Heartfelt thanks are due David Groff for his superb editorial assistance, and to Frank Weimann of The Literary Group in New York for taking the manuscript, with all his considerable skill, to the publishers. The assistance of Amir Fedder and Rich Capogrosso at The Artists Agency, and Jessica Wainwright, Lauren Mactas, and Kim Marsar at The Literary Group was also invaluable and much appreciated.

I think it wise for me to also thank the guardian angels that I have overworked over the years because they probably had something to do with my getting Tom Spain of Delacorte Press as my editor. Tom's ability to find the core truths of this book (and sometimes even pointing them out to me when I didn't see them) has shaped it into all that it is. Many thanks to Delacorte's Mitch Hoffman for his many kindnesses. Karen Mender, Carisa Hays, Linda Steinman, and Vicki Flick have also been very helpful. Throughout this whole process, Carole Baron, Leslie Schnur, and all of their staff at Delacorte have been unfailingly enthusiastic and professional in their support of this book. I very much appreciate their efforts.

I wish to express my appreciation to Chuck Gordon, Mark Sternberg, and Peter Cramer of Daybreak Productions for taking what was then an uncompleted manuscript and working it through the Hollywood labyrinth, and to Joe Johnston, Larry Franco, Lewis Colick, and all of the staff of the Universal Studios motion picture production crew for translating my story to the silver screen. They made the impossible happen.

I cannot thank enough my readers of the manuscript as it developed, especially Linda Terry, who saw it from the first very rough drafts and helped me to improve it through all the versions that followed. My thanks further to Linda for her love and support during the entire period of creation. I could not have done it without her. Much help also came from Big Creek High School classmate Emily Sue Buckberry, who kindly offered historical corrections, editorial commentary, and morale boosts throughout. Special thanks are due Harry Kenneth Lavender, my uncle, for his technical assistance on coal mining and life in general in the coalfields.

Perry Turner and Pat Trenner, editors at *Air & Space/Smithsonian* magazine, are due many thanks for publishing the article, "The Big Creek Missile Agency," which gained the attention that led to this book.

Finally, much gratitude is owed to the men who were once the rocket boys for agreeing to let me write about them as they were so many years ago, to Mrs. Jan Siers, who gave me permission to include Sherman, to my brother Jim, who assented to my dredging up our teenaged conflicts, to several high-school classmates who wish to remain anonymous but assisted me and are in this book in one guise or another, and to my mother, who has maintained her sense of humor over the sometimes strange maneuvers of her second son to this day.

—Homer H. Hickam, Jr.
May 1998

CONTENTS

ROCKET BOYS

All one can really leave one's children is what's inside their heads. Education, in other words, and not earthly possessions, is the ultimate legacy, the only thing that cannot be taken away.

—Dr. Wernher von Braun

All I've done is give you a book. You have to have the courage to learn what's inside it.

—Miss Freida Joy Riley

1

COALWOOD

UNTIL I BEGAN to build and launch rockets, I didn't know my hometown was at war with itself over its children and that my parents were locked in a kind of bloodless combat over how my brother and I would live our lives. I didn't know that if a girl broke your heart, another girl, virtuous at least in spirit, could mend it on the same night. And I didn't know that the enthalpy decrease in a converging passage could be transformed into jet kinetic energy if a divergent passage was added. The other boys discovered their own truths when we built our rockets, but those were mine.

Coalwood, West Virginia, where I grew up, was built for the purpose of extracting the millions of tons of rich, bituminous coal that lay beneath it. In 1957, when I was fourteen years old and first began to build my rockets, there were nearly two thousand people living in Coalwood. My father, Homer Hickam, was the mine superintendent, and our house was situated just a few hundred yards from the mine's entrance, a vertical shaft eight hundred feet deep. From the window of my bedroom, I could see the black steel tower that sat over the shaft and the comings and goings of the men who worked at the mine.

Another shaft, with railroad tracks leading up to it, was used to bring out the coal. The structure for lifting, sorting, and dumping the coal was called the tipple. Every weekday, and even on Saturday when times were good, I could watch the black coal cars rolling beneath the tipple to receive their massive loads and then smoke-spouting locomotives straining to pull them away. All through the day, the heavy thump of the locomotives' steam pistons thundered down our narrow valleys, the town shaking to the crescendo of grinding steel as the great trains accelerated. Clouds of coal dust rose from the open cars, invading everything, seeping through windows and creeping under doors. Throughout my childhood, when I raised my blanket in the morning, I saw a black, sparkling powder float off it. My socks were always black with coal dirt when I took my shoes off at night.

Our house, like every house in Coalwood, was company-owned. The company charged a small monthly rent, automatically deducted from the miners' pay. Some of the houses were tiny and single-storied, with only one or two bedrooms. Others were big two-story duplexes, built as boardinghouses for bachelor miners in the booming 1920's and later sectioned off as individual family dwellings during the Depression. Every five years, all the houses in Coalwood were painted a company white, which the blowing coal soon tinged gray. Usually in the spring, each family took it upon themselves to scrub the exterior of their house with hoses and brushes.

Each house in Coalwood had a fenced-off square of yard. My mother, having a larger yard than most to work with, planted a rose garden. She hauled in dirt from the mountains by the sackful, slung over her shoulder, and fertilized, watered, and manicured each bush with exceeding care. During the spring and summer, she was rewarded with bushes filled with great blood-red blossoms as well as dainty pink and yellow buds, spatters of brave color against the dense green of the heavy forests that surrounded us and the gloom of the black and gray mine just up the road.

Our house was on a corner where the state highway turned east toward the mine. A company-paved road went the other way to the center of town. Main Street, as it was called, ran down a valley so narrow in places that a boy with a good arm could throw a rock from one side of it to the other. Every day for the three years before I went to high school, I got on my bicycle in the morning with a big white canvas bag strapped over my shoulder and deliv-

ered the *Bluefield Daily Telegraph* down this valley, pedaling past the
Coalwood School and the rows of houses that were set along a
little creek and up on the sides of the facing mountains. A mile
down Main was a large hollow in the mountains, formed where
two creeks intersected. Here were the company offices and also the
company church, a company hotel called the Club House, the post
office building, which also housed the company doctor and the
company dentist, and the main company store (which everybody
called the Big Store). On an overlooking hill was the turreted mansion occupied by the company general superintendent, a man sent
down by our owners in Ohio to keep an eye on their assets. Main
Street continued westward between two mountains, leading to clus-
ters of miners' houses we called Middletown and Frog Level. Two
forks led up mountain hollows to the "colored" camps of Mudhole
and Snakeroot. There the pavement ended, and rutted dirt roads
began.

At the entrance to Mudhole was a tiny wooden church presided
over by the Reverend "Little" Richard. He was dubbed "Little"
because of his resemblance to the soul singer. Nobody up Mudhole
Hollow subscribed to the paper, but whenever I had an extra one, I
always left it at the little church, and over the years, the Reverend
Richard and I became friends. I loved it when he had a moment to
come out on the church porch and tell me a quick Bible story while
I listened, astride my bike, fascinated by his sonorous voice. I
especially admired his description of Daniel in the lions' den. When
he acted out with bug-eyed astonishment the moment Daniel's
captors looked down and saw their prisoner lounging around in the
pit with his arm around the head of a big lion, I laughed apprecia-
tively. "That Daniel, he knew the Lord," the Reverend summed up
with a chuckle while I continued to giggle, "and it made him brave.
How about you, Sonny? Do you know the Lord?"

I had to admit I wasn't certain about that, but the Reverend said
it was all right. "God looks after fools and drunks," he said with a
big grin that showed off his gold front tooth, "and I guess he'll
look after you too, Sonny Hickam." Many a time in the days to
come, when I was in trouble, I would think of Reverend Richard

and his belief in God's sense of humor and His fondness for ne'er-do-wells. It didn't make me as brave as old Daniel, but it always gave me at least a little hope the Lord would let me scrape by.

The company church, the one most of the white people in town went to, was set down on a little grassy knob. In the late 1950's, it came to be presided over by a company employee, Reverend Josiah Lanier, who also happened to be a Methodist. The denomination of the preacher the company hired automatically became ours too. Before we became Methodists, I remember being a Baptist and, once for a year, some kind of Pentecostal. The Pentecostal preacher scared the women, hurling fire and brimstone and warnings of death from his pulpit. When his contract expired, we got Reverend Lanier.

I was proud to live in Coalwood. According to the West Virginia history books, no one had ever lived in the valleys and hills of McDowell County before we came to dig out the coal. Up until the early nineteenth century, Cherokee tribes occasionally hunted in the area, but found the terrain otherwise too rugged and uninviting. Once, when I was eight years old, I found a stone arrowhead embedded in the stump of an ancient oak tree up on the mountain behind my house. My mother said a deer must have been lucky some long ago day. I was so inspired by my find that I invented an Indian tribe, the Coalhicans, and convinced the boys I played with—Roy Lee, O'Dell, Tony, and Sherman—that it had really existed. They joined me in streaking our faces with berry juice and sticking chicken feathers in our hair. For days afterward, our little tribe of savages formed raiding parties and conducted massacres throughout Coalwood. We surrounded the Club House and, with birch-branch bows and invisible arrows, picked off the single miners who lived there as they came in from work. To indulge us, some of them even fell down and writhed convincingly on the Club House's vast, manicured lawn. When we set up an ambush at the tipple gate, the miners going on shift got into the spirit of things, whooping and returning our imaginary fire. My father observed this from his office by the tipple and came out to restore order. Although the Coalhicans escaped into the hills, their chief was reminded at the supper table that night that the mine was for work, not play.

When we ambushed some older boys—my brother, Jim, among them—who were playing cowboys up in the mountains, a great mock battle ensued until Tony, up in a tree for a better line of sight, stepped on a rotted branch and fell and broke his arm. I organized the construction of a litter out of branches, and we bore the great warrior home. The company doctor, "Doc" Lassiter, drove to Tony's house in his ancient Packard and came inside. When he caught sight of us still in our feathers and war paint, Doc said he was the "heap big medicine man." Doc set Tony's arm and put it in a cast. I remember still what I wrote on it: *Tony—next time pick a better tree.* Tony's Italian immigrant father was killed in the mine that same year. He and his mother left and we never heard from them again. This did not seem unusual to me: A Coalwood family required a father, one who worked for the company. The company and Coalwood were one and the same.

I learned most of what I knew about Coalwood history and my parents' early years at the kitchen table after the supper dishes were cleared. That was when Mom had herself a cup of coffee and Dad a glass of milk, and if they weren't arguing about one thing or the other, they would talk about the town and the people in it, what was going on at the mine, what had been said at the last Women's Club meeting, and, sometimes, little stories about how things used to be. Brother Jim usually got bored and asked to be excused, but I always stayed, fascinated by their tales.

Mr. George L. Carter, the founder of Coalwood, came in on the back of a mule in 1887, finding nothing but wilderness and, after he dug a little, one of the richest seams of bituminous coal in the world. Seeking his fortune, Mr. Carter bought the land from its absentee owners and began construction of a mine. He also built houses, school buildings, churches, a company store, a bakery, and an icehouse. He hired a doctor and a dentist and provided their services to his miners and their families for free. As the years passed and his coal company prospered, Mr. Carter had concrete sidewalks poured, the streets paved, and the town fenced to keep cows from roaming the streets. Mr. Carter wanted his miners to have a decent place to live. But in return, he asked for a decent day's work. Coalwood was, after all, a place for work above all else: hard, bruising, filthy, and sometimes deadly work.

When Mr. Carter's son came home from World War I, he brought with him his army commander, a Stanford University graduate of great engineering and social brilliance named William Laird, who everyone in town called, with the greatest respect and deference, the Captain. The Captain, a big expansive man who stood nearly six and a half feet tall, saw Coalwood as a laboratory for his ideas, a place where the company could bring peace, prosperity, and tranquillity to its citizens. From the moment Mr. Carter hired him and placed him in charge of operations, the Captain began to implement the latest in mining technology. Shafts were sunk for ventilation, and as soon as it was practical, the mules used to haul out the coal from the mine were replaced by electric motors. Later, the Captain stopped all the hand digging and brought in giant machines, called continuous miners, to tear the coal from its seams. The Captain expanded Mr. Carter's building program, providing every Coalwood miner a house with indoor plumbing, a Warm Morning stove in the living room, and a coal box the company kept full. For the town's water supply, he tapped into a pristine ancient lake that lay a thousand feet below. He built parks on both ends of the town and funded the Boy Scouts, Girl Scouts, Brownies, Cub Scouts, and the Women's Club. He stocked the Coalwood school library and built a school playground and a football field. Because the mountains interfered with reception, in 1954 he erected an antenna on a high ridge and provided one of the first cable television systems in the United States as a free service.

Although it wasn't perfect, and there was always tension between the miners and the company, mostly about pay, Coalwood was, for a time, spared much of the violence, poverty, and pain of the other towns in southern West Virginia. I remember sitting on the stairs in the dark listening to my father's father—my Poppy— talk to Dad in our living room about "bloody Mingo," a county just up the road from us. Poppy had worked there for a time until a war broke out between union miners and company "detectives." Dozens of people were killed and hundreds were wounded in pitched battles with machine guns, pistols, and rifles. To get away from the violence, Poppy moved his family first to Harlan County, Kentucky, and then, when battles erupted there, to McDowell County, where he went to work in the Gary mine. It was an improvement, but Gary was still a place of strikes and lockouts and the occasional bloody head.

In 1934, when he was twenty-two years old, my father applied for work as a common miner with Mr. Carter's company. He came because he had heard that a man could make a good life for himself in Coalwood. Almost immediately, the Captain saw something in the skinny, hungry lad from Gary—some spark of raw intelligence, perhaps—and took him as a protégé. After a couple of years, the Captain raised Dad to section foreman, taught him how to lead men and operate and ventilate a mine, and instilled in him a vision of the town.

After Dad became a foreman, he convinced his father to quit the Gary mine and move to Coalwood, where there was no union and a man could work. He also wrote Elsie Lavender, a Gary High School classmate who had moved on her own to Florida, to come back to West Virginia and marry him. She refused. Whenever the story was told, Mom took over at this point and said the letter she next received was from the Captain, who told her how much Dad loved her and needed her, and would she please stop being so stubborn down there in the palm trees and come to Coalwood and marry the boy? She agreed to come to Coalwood to visit, and one night at the movies in Welch, when Dad asked her to marry him again, she said if he had a Brown Mule chewing tobacco wrapper in his pocket, she'd do it. He had one and she said yes. It was a decision that I believed she often regretted, but still would not have changed.

Poppy worked in the Coalwood mine until 1943, when a runaway mine car cut off both his legs at the hip. He spent the rest of his life in a chair. My mother said that after the accident, Poppy was in continuous pain. To take his mind off it, he read nearly every book in the County Library in Welch. Mom said when she and Dad visited him, Poppy would be hurting so much he could hardly talk, and Dad would agonize over it for days afterward. Finally, a doctor prescribed paregoric, and as long as he had a continuous supply, Poppy found some peace. Dad saw that Poppy had all the paregoric he wanted. Mom said after the paregoric, Poppy never read another book.

Because he was so dedicated to the Captain and the company, I saw little of my father while I was growing up. He was always at the mine, or sleeping prior to going to the mine, or resting after getting back. In 1950, when he was thirty-eight years old, he developed cancer of the colon. At the time, he was working double shifts, leading a section deep inside the mine charged with cutting through a massive rock header. Behind the dense sandstone of the header, the Captain believed, was a vast, undiscovered coal seam. Nothing was more important to my father than to get through the header and prove the Captain right. After months of ignoring the bloody symptoms of his cancer, Dad finally passed out in the mine. His men had to carry him out. It was the Captain, not my mother, who rode with him in the ambulance to the hospital in Welch. There the doctors gave him little chance for survival. While Mom waited in the Stevens Clinic waiting room, the Captain was allowed to watch the operation. After a long piece of his intestine was removed, Dad confounded everybody by going back to work in a month. Another month later, drenched in rock dust and sweat, his section punched through the header into the softest, blackest, purest coal anyone had ever seen. There was no celebration. Dad came home, showered and scrubbed himself clean, and went to bed for two days. Then he got up and went back to work again.

There were at least a few times the family was all together. When I was little, Saturday nights were reserved for us to journey over to the county seat of Welch, seven miles and a mountain away from Coalwood. Welch was a bustling little commercial town set down by the Tug Fork River, its tilted streets filled with throngs of miners and their families come to shop. Women went from store to store with children in their arms or hanging from their hands, while their men, often still in mine coveralls and helmets, lagged behind to talk about mining and high-school football with their fellows.

While Mom and Dad visited the stores, Jim and I were deposited at the Pocahontas Theater to watch cowboy movies and adventure serials with hundreds of other miners' kids. Jim would never talk to any of the others, but I always did, finding out where the boy or girl who sat next to me was from. It always seemed exciting to me when I met somebody from exotic places like Keystone or Iaeger, mining towns on the other side of the county. By the time I had visited and then watched a serial and a double feature and then been retrieved by my parents to walk around Welch to finish up Mom's shopping, I was exhausted. I almost always fell sound asleep on the ride home in the backseat of the car. When we got back to Coalwood, Dad would lift me over his shoulder and carry me to bed. Sometimes even when I wasn't asleep I pretended to be, just to know his touch.

Shift changes in Coalwood were daily major events. Before each shift began, the miners going to work came out of their houses and headed toward the tipple. The miners coming off-shift, black with coal dirt and sweat, formed another line going in the opposite direction. Every Monday through Friday, the lines formed and met at intersections until hundreds of miners filled our streets. In their coveralls and helmets, they reminded me of newsreels I'd seen of soldiers slogging off to the front.

Like everybody else in Coalwood, I lived according to the rhythms set by the shifts. I was awakened in the morning by the tromp of feet and the clunking of lunch buckets outside as the day shift went to work, I ate supper after Dad saw the evening shift down the shaft, and I went to sleep to the ringing of a hammer on steel and the dry hiss of an arc welder at the little tipple machine shop during the hoot-owl shift. Sometimes, when we boys were still in grade school and tired of playing in the mountains, or dodgeball by the old garages, or straight base in the tiny clearing behind my house, we would pretend to be miners ourselves and join the men in their trek to the tipple. We stood apart in a knot and watched them strap on their lamps and gather their tools, and then a bell would ring, a warning to get in the cage. After they were swallowed by the earth, everything became eerily quiet. It was an unsettling moment, and we boys were always glad to get back to our games, yelling and brawling a little louder than necessary to shatter the spell cast on us by the tipple.

Coalwood was surrounded by forests and mountains dotted with caves and cliffs and gas wells and fire towers and abandoned mines just waiting to be discovered and rediscovered by me and the boys and girls I grew up with. Although our mothers forbade it, we also played around the railroad tracks. Every so often, somebody would come up with the idea of putting a penny on the track and getting it run over by the coal cars to make a big flat medal. We'd all do it then until we had used up our meager supply. Stifling our laughter, we'd hand the crushed coppers across the counter at the company store for candy. The clerk, having seen this many times over the years, usually accepted our tender without comment. They probably had a stack of flat pennies somewhere in the company-store offices, collected over the decades.

For a satisfying noise, nothing beat going up on the Coalwood School bridge and throwing pop bottles into the empty coal cars rolling in to the tipple. When the coal cars were full and stopped beneath the bridge, some of the braver boys would even leap into them, plunging waist-deep into the loose coal. I tried it once and barely escaped when the train suddenly pulled out, bound for Ohio. I wallowed through the coal and climbed down the outside ladder of the car and jumped for it, skinning my hands, knees, and elbows on the packed coal around the track. My mother took no pity on me and scrubbed the coal dirt off me with a stiff brush and Lava soap. My skin felt raw for a week.

When I wasn't outside playing, I spent hours happily reading. I loved to read, probably the result of the unique education I received from the Coalwood School teachers known as the "Great Six," a corruption of the phrase "grades one through six." For years, these same six teachers had seen through their classrooms generations of Coalwood students. Although Mr. Likens, the Coalwood School principal, controlled the junior high school with a firm hand, the Great Six held sway in the grades below. It seemed to be very important to these teachers that I read. By the second grade, I was intimately familiar with and capable of discussing in some detail *Tom Sawyer* and *Uncle Tom's Cabin*. *Huckleberry Finn* they saved for me until the third grade, tantalizingly holding it back as if it contained the very secrets of life. When I was finally allowed to read it, I very well knew this was no simple tale of rafting down a river but the everlasting story of America itself, with all our glory and shame.

Bookcases filled with complete sets of *Tom Swift*, *The Bobbsey Twins*, *The Hardy Boys*, and *Nancy Drew* were in the grade-school hallway and available to any student for the asking. I devoured them, savoring the adventures they brought to me. When I was in the fourth grade, I started going upstairs to the junior high school library to check out the *Black Stallion* series. There, I also discovered Jules Verne. I fell in love with his books, filled as they were with not only great adventures but scientists and engineers who considered the acquisition of knowledge to be the greatest pursuit of mankind. When I finished all the Verne books in the library, I became the first in line for any book that arrived written by modern science-fiction writers such as Heinlein, Asimov, van Vogt, Clarke, and Bradbury. I liked them all unless they branched out into fantasy. I didn't care to read about heroes who could read minds or walk through walls or do magic. The heroes I liked had courage and knew more real stuff than those who opposed them. When the Great Six inspected my library record and found it top-heavy with adventure and science fiction, they prescribed appropriate doses of Steinbeck, Faulkner, and F. Scott Fitzgerald. It seemed as if all through grade school, I was reading two books, one for me and one for my teachers.

For all the knowledge and pleasure they gave me, the books I read in childhood did not allow me to see myself past Coalwood. Almost all the grown-up Coalwood boys I knew had either joined the military services or gone to work in the mine. I had no idea what the future held in store for me. The only thing I knew for sure was my mother did not see me going into the mine. One time after Dad tossed her his check, I heard her tell him, "Whatever you make, Homer, it isn't enough."

He replied, "It keeps a roof over your head."

She looked at the check and then folded it and put it in her apron pocket. "If you'd stop working in that hole," she said, "I'd live under a tree."

After Mr. Carter sold out, the company was renamed Olga Coal Company. Mom always called it "Miss Olga." If anybody asked her where Dad was, she'd say, "With Miss Olga." She made it sound as if it was his mistress.

Mom's family did not share her aversion to coal mining. All of her four brothers—Robert, Ken, Charlie, and Joe—were miners, and her sister, Mary, was the wife of a miner. Despite their father's hideous accident, my father's two brothers were also miners; Clarence worked in the Caretta mine across the mountain from Coalwood, and Emmett in mines around the county. Dad's sister, Bennie, married a Coalwood miner and they lived down across the creek, near the big machine shops. But the fact that all of her family, and my father's family, were miners did not impress my mother. She had her own opinion, formed perhaps by her independent nature or by her ability to see things as they really were, not as others, including herself, would wish them to be.

In the morning before she began her ritual battle against the dust, my mother could nearly always be found with a cup of coffee at the kitchen table in front of an unfinished mural of a seashore. She had been working on the painting ever since Dad took over the mine and we moved into the Captain's house. By the fall of 1957, she had painted in the sand and shells and much of the sky and a couple of seagulls. There was an indication of a palm tree going up too. It was as if she was painting herself another reality. From her seat at the table, she could reflect on her roses and bird feeders through the picture window the company carpenters had installed for her. Per her specifications, it was angled so not a hint of the mine could be seen.

I knew, even as a child, that my mother was different from just about everybody in Coalwood. When I was around three years old, we were visiting Poppy in his little house up Warriormine Hollow, and he took me on his lap. That scared me, because he didn't have a lap, just an empty wrinkled blanket where his legs should have been. I struggled in his thick arms while Mom hovered nervously nearby. "He's just like Homer," I remember toothless Poppy lisping to Mom while I squirmed. He called to my dad on the other side of the room. "Homer, he's just like you!"

Mom anxiously took me from Poppy and I clutched hard to her shoulder, my heart beating wildly from an unidentified terror. She carried me out onto the front porch, stroking my hair and hushing me. "No, you're not," she crooned just loud enough so only she and I could hear. "No, you're not."

Dad slapped open the screen door and came out on the porch as if to argue with her. Mom turned away from him and I saw his eyes, usually a bright hard blue, soften into liquid blots. I snuggled my face into her neck while Mom continued to rock and hold me, still singing her quietly insistent song: *No, you're not. No, you're not.* All through my growing-up years, she kept singing it, one way or the other. It was only when I was in high school and began to build my rockets that I finally understood why.

2

SPUTNIK

I WAS ELEVEN years old when the Captain retired and my father
took his position. The Captain's house, a big, barnlike wood-frame
structure, and the closest house in Coalwood to the tipple, became
our house. I liked the move because for the first time I didn't have
to share a room with Jim, who never made any pretense of liking
me or wanting me around. From my earliest memory, it was clear
my brother blamed me for the tension that always seemed to exist
between our parents. There may have been a kernel of truth to his
charge. The story I heard from Mom was that Dad wanted a
daughter, and when I came along he was so clearly disappointed,
and said so in such certain terms, she retaliated by naming me after
him: Homer Hadley Hickam, *Junior.* Whether that incident caused
all their other arguments that followed, I couldn't say. All I knew
was that their discontent had left me with a heavy name. Fortu-
nately, Mom started calling me "Sunny" right away because, she
said, I was a happy child. So did everybody else, although my first-
grade teacher changed the spelling to the more masculine "Sonny."

Mr. McDuff, the mine carpenter, built me a desk and some
bookshelves for my new room, and I stocked them with science-
fiction books and model airplanes. I could happily spend hours
alone in my room.

In the fall of 1957, after nine years of classes in the Coalwood
School, I went across the mountains to Big Creek, the district high
school, for the tenth through the twelfth grade. Except for having

to get up to catch the school bus at six-thirty in the morning, I liked high school right off. There were kids there from all the little towns in the district, and I started making lots of new friends, although my core group remained my buddies from Coalwood: Roy Lee, Sherman, and O'Dell.

I guess it's fair to say there were two distinct phases to my life in West Virginia: everything that happened before October 5, 1957 and everything that happened afterward. My mother woke me early that morning, a Saturday, and said I had better get downstairs and listen to the radio. "What is it?" I mumbled from beneath the warm covers. High in the mountains, Coalwood could be a damp, cold place even in the early fall, and I would have been happy to stay there for another couple of hours, at least.

"Come listen," she said with some urgency in her voice. I peeked at her from beneath the covers. One look at her worried frown and I knew I'd better do what she said, and fast.

I threw on my clothes and went downstairs to the kitchen, where hot chocolate and buttered toast waited for me on the counter. There was only one radio station we could pick up in the morning, WELC in Welch. Usually, the only thing WELC played that early was one record dedication after the other for us high-school kids. Jim, a year ahead of me and a football star, usually got several dedications every day from admiring girls. But instead of rock and roll, what I heard on the radio was a steady *beep-beep-beep* sound. Then the announcer said the tone was coming from something called *Sputnik*. It was Russian and it was in space. Mom looked from the radio to me. "What is this thing, Sonny?"

I knew exactly what it was. All the science-fiction books and Dad's magazines I'd read over the years put me in good stead to answer. "It's a space satellite," I explained. "We were supposed to launch one this year too. I can't believe the Russians beat us to it!"

She looked at me over the rim of her coffee cup. "What does it do?"

"It orbits around the world. Like the moon, only closer. It's got science stuff in it, measures things like how cold or hot it is in space. That's what ours was supposed to do, anyway."

"Will it fly over America?"

I wasn't certain about that. "I guess," I said.

Mom shook her head. "If it does, it's going to upset your dad, no end."

I knew that was the truth. As rock-ribbed a Republican as ever was allowed to take a breath in West Virginia, my father detested the Russian Communists, although, it should be said, not quite as much as certain American politicians. For Dad, Franklin Delano Roosevelt was the Antichrist, Harry Truman the vice-Antichrist, and UMWA chief John L. Lewis was Lucifer himself. I'd heard Dad list all their deficiencies as human beings whenever my Uncle Ken—Mom's brother—came to visit. Uncle Ken was a big Democrat, like his father. Uncle Ken said his daddy would've voted for our dog Dandy before he'd have voted for a Republican. Dad said he'd do the same before casting a ballot for a Democrat. Dandy was a pretty popular politician at our house.

All day Saturday, the radio announcements continued about the Russian *Sputnik*. It seemed like each time there was news, the announcer was more excited and worried about it. There was some talk as to whether there were cameras on board, looking down at the United States, and I heard one newscaster wonder out loud if maybe an atomic bomb might be aboard. Dad was working at the mine all day, so I didn't get to hear his opinion on what was happening. I was already in bed by the time he got home, and on Sunday, he was up and gone to the mine before the sun was up. According to Mom, there was some kind of problem with one of the continuous miners. Some big rock had fallen on it. At church, Reverend Lanier had nothing to say about the Russians or *Sputnik* during his sermon. Talk on the church steps afterward was mostly about the football team and its undefeated season. It was taking awhile for *Sputnik* to sink in, at least in Coalwood.

By Monday morning, almost every word on the radio was about *Sputnik*. Johnny Villani kept playing the beeping sound over and over. He talked directly to students "across McDowell County" about how we'd better study harder to "catch up with the Russians." It seemed as if he thought if he played us his usual rock and roll, we might get even farther behind the Russian kids. While I listened to the beeping, I had this mental image of Russian high-school kids lifting the *Sputnik* and putting it in place on top of a big, sleek rocket. I envied them and wondered how it was they were so smart. "I figure you've got about five minutes or you're going to miss your bus," Mom pointed out, breaking my thinking spell.

I gulped down my hot chocolate and dashed up the steps past Jim coming down. Not surprisingly, Jim had every golden hair on his head in place, the peroxide curl in front just so, the result of an hour of careful primping in front of the medicine-chest mirror in the only bathroom in the house. He was wearing his green and white football letter jacket and also a new button-down pink and black shirt (collar turned up), pegged chino pants with a buckle in the back, polished penny loafers, and pink socks. Jim was the best-dressed boy in school. One time when Mom got Jim's bills from the men's stores in Welch, she said my brother must have been dropped off by mistake by vacationing Rockefellers. In contrast, I was wearing a plaid flannel shirt, the same pair of cotton pants I'd worn to school all the previous week, and scuffed leather shoes, the ones I'd worn the day before playing around the creek behind the house. Jim and I said nothing as we passed on the stairs. There was nothing to say. I would tell people some years later that I was raised an only child and so was my brother.

This is not to say Jim and I didn't have a history. From the first day I could remember being alive, he and I had brawled. Although I was smaller, I was sneakier, and we had battled so many times over the years that I knew all his moves, knew that as long as I kept inside the swing of his fists, he wasn't going to kill me. By the fall of 1957, Jim and I were about two months into a period of uneasy truce. Our last fight had scared us both into it. It began when Jim found my bike lying on top of his in the backyard. My bike's kickstand had collapsed (I probably hadn't levered it all the way down) and my bike had fallen on top of his, taking them both down. Furious, he carried my bike to the creek and threw it in. Mom was over in Welch shopping and Dad was at the mine. Jim stomped up to my room where I was lounging on my bed reading a book, slammed open the door, and told me what he had done and why. "If anything of yours ever touches anything of mine again," he bellowed, "I'll beat the ever-loving hell out of you!"

"How about right now, fat boy?" I cried, launching myself at him. We fell into the hall, me on the inside punching him in the stomach and him yowling and swinging at the air until we rolled down the stairs and crashed into the foyer, where I managed a lucky hit to his ear with my elbow. Howling, he picked me up and hurled me into the dining room, but I got right up and hit him with one of Mom's prized cherry-wood chairs, breaking off one of its legs. He chased me into the kitchen, whereupon I picked up a metal pot off the stove and bounced it off his noggin. Then I made for the back porch, but he tackled me and we fell through the screen door, ripping it off its hinges. We wrestled in the grass until he got up and then leapt back on top of me. That's when I felt my ribs creak. My chest hurt so bad I started to cry, but I didn't say anything mainly because I couldn't breathe. His leg was in my face, so I bit him as hard as I could to make him get off me. He screamed and jumped up while I rolled over onto my back and gasped for air. My ribs felt like they were caved in. Blood flowed from my nose. A knot on Jim's head was rising, and there was going to be a nice, purple welt on his leg. We had managed some real damage to each other and knew we'd gone too far at last.

When Mom came home, she found our bikes parked neatly beside each other in the backyard and Jim and I sitting innocently together in the living room. Jim had his hand on his head, idly cradling it while he read the sports page of the *Welch Daily News*. I was sitting nearby, watching television, trying not to scream from the pain each time I breathed. My ribs ached for a month. The dining-room chair was back in its place, well-glued. Jim and I watched it for days to keep anyone from sitting on it until it dried. The dogs got the blame for the screen door. Either Mom never noticed it or chose not to mention the dent in her pot.

Jim was already at the bus stop while I was still rushing around getting ready. I was in and out of the bathroom in two minutes flat, pausing only to brush my teeth and run a wet hand through my hair. I had my mother's hair—black, thick, and curly. She had started turning gray in her thirties, so I knew that was likely to happen to me too. It didn't look like I'd gotten anything from Dad's family tree. Mom said I was a Lavender like her through and

through. Dad never argued with her about it, so I guessed it was
so. That was fine with me. The Hickams always seemed a nervous
bunch to me. Dad and his brother Clarence and sister Bennie never
seemed able to quite settle down, always jumping up to walk real
fast to wherever they were going, and talking fast too. The
Lavenders were a more relaxed bunch, although Mom's father, my
"Ground-Daddy," was shot in the arm crawling into some lady's
bedroom while her husband was supposed to be working the hoot-
owl shift over in Gary. My mom said her mama helped her daddy
put his coat on while his winged arm healed. Mom also said
Ground-Daddy would have gone naked out in the snow before *she*
would've helped him.

On the first school day after *Sputnik*, I threw on one of Jim's
hand-me-down cotton jackets, grabbed my books off the banister,
and snatched the brown-bag lunch Mom held out for me at the
front door. I had to run for it. The big yellow bus was already at
the stop in front of the Todds' house, and Jack Martin, the driver,
waited for no one. He watched sourly, an unlit cigar clamped between his teeth, as I scrambled aboard an inch ahead of the closing
doors. "Any later and you'd be walking, Sonny boy," he said. I
knew he wasn't kidding. Jack ran his bus in dictatorial fashion. The
slightest breach in decorum would find the perpetrator kicked off
on the side of the road, no matter where we were. I found a two-
inch sliver of a seat on a bench and squeezed in beside Linda
DeHaven and Margie Jones, girls who had been in my class since
the first grade. They shifted minutely and fell back asleep. Jack
changed gears and we were off. My friend and fellow former
Coalhican O'Dell was snoozing up front, just behind Jack. O'Dell
was small and excitable. His hair was the pale, nearly translucent
color of spun silk. In the seat behind him, Sherman, a compact,
muscular kid with a wide, intelligent face, was also sleeping. Sherman's left leg was shriveled and weak, the result of polio. During
all our years growing up together, he never complained about his
affliction and I never gave it any mind. He either kept up with the
rest of us or he didn't.

Roy Lee, thin and long-legged, got on the bus at the next stop,
easing down the aisle until he squeezed in behind me. For as long

as I could remember, Roy Lee and I had been friends. He'd show up at my house or I'd go up to his and we'd be off to the mountains, playing cowboys or spacemen or pirates or whatever we could think up. Roy Lee was unique among us. He had his own car, the result of an insurance settlement after his father had died in the mine. His mother, wanting to keep Roy Lee in Coalwood, had campaigned to keep her company house. Surprisingly, she and Roy Lee had been allowed to stay. Maybe it was because Roy Lee's brother still worked in the mine. Roy Lee was a good-looking kid, and he knew it. His hair was coal black, and he kept it swept back and greased and teased into what we called a D.A. (duck's ass). He looked a bit like a very young Elvis. Roy Lee thought he was pretty much girl bait, and I guess he was, seeing as how he had a date nearly every weekend. Owning a car probably helped too.

I was grateful to have Roy Lee, Sherman, and O'Dell as friends. When I entered the first grade, I found myself in a community of boys from all over the town, and it became apparent that, as my father's son, I was marked by his position. Around the kitchen tables at night, union fathers often identified Homer Hickam as the enemy, and the boys from those families sometimes went out looking for revenge. Jim was always big for his age and known for his terrible temper. I was a much easier target, caught at recess behind the school or loitering around the Big Store. Though I came home bloody, I never told my mother who attacked me, and my father never knew of it at all. Coalwood boys didn't carry tales on one another. I did the best I could for a small, nearsighted kid and each year got to be a little tougher nut to crack. I even managed to bloody a few noses myself. For some reason, Roy Lee, Sherman, and O'Dell never seemed to mind who my dad was. As far as they were concerned, we were all just Coalwood kids together.

The road out of town led past the coal mine, and Jack blew the bus horn at the tipple. Those of us still awake waved at the men at the man-hoist, and then we kept going for about a mile until we stopped for the few students that came out of the hollow at Six (named after the sixth ventilation shaft sunk for the mine; there were some houses built up around it). They were the last students to pick up. Then we started up the first of the mountains. Between

Coalwood and Big Creek High School were eight miles of twisting, potholed roads. Unless it was snowing, it took Jack about forty-five minutes to cover the distance.

The road up Coalwood Mountain turned through one steeply inclined switchback after another. Wedged three to a seat, most of us dozed, leaning against one another at each turn. At the top of the mountain, the road dropped precipitously and swung back and forth until it bottomed out into a long, narrow valley. Here was the longest stretch of straight road in the district, nearly a mile of asphalt. About midway down it, behind a barbed-wire fence, was one of the big fans that ventilated the mine. On Saturday nights, this straight stretch—nicknamed Little Daytona—was a racetrack for those few teenagers with wheels, and the fan a favorite place to park and make out. Since I had neither a driver's license or a girlfriend, I knew both those things only by hearsay. Roy Lee was my most likely source. He had told me he took his dates to park there after going to the Dugout. The Dugout was in the basement of the Owl's Nest restaurant across from the high school, and there were dances there every Saturday night. I'd never been to the Dugout, but from what I'd heard it was a lot of fun. One of the Big Creek janitors, Ed Johnson, was the disc jockey, and Roy Lee said he had one of the best record collections this side of *American Bandstand.*

After a sharp turn at the end of Little Daytona, we entered the town of Caretta. Caretta was owned by the same company that owned Coalwood. Its tunnels had broken through to our mine the previous year. There had been a massive slab of sandstone between the two mines, and my father had fought through it like he was in a war. Once opened, the combined mines caused so many ventilation problems Dad had to take over both of them. According to what I heard Mom tell Uncle Joe during a visit, a lot of people in Caretta had said some real nasty things about that, calling Dad "uppity." There seemed to be so many people that just couldn't forgive Dad for not having a college degree like the Captain. That seemed strange to me since they didn't have a degree either. Mom told Uncle Joe that, as far as she was concerned, those Caretta people weren't "much punkin' and funny turned too." Mom sometimes seemed to lapse into a different dialect when one of her brothers was around. I remember Uncle Joe nodding his head in solemn agreement.

After we passed through Caretta, we reached a fork in the road at a little place called Premier, where there was an old run-down whitewashed brick building called the Spaghetti House. I'd never been in there, but Roy Lee had. He said there were whores in there, old skaggy ones that would give you the clap. I didn't know what the clap was, but it didn't sound like I wanted it. Roy Lee said he'd only been in there one time to get change for a dollar and they had given him four rubbers instead. He still had all four. I knew because he'd shown them to me. He carried one of them in his billfold. It looked pretty old to me.

War Mountain was not as steep as the mountain out of Coalwood, but its roads were narrower and there were two curves that nearly doubled back on themselves. Jack slowed down to a crawl at each of these, blowing the bus horn and then easing us around. Those of us pinned to the outer side of the bus looked straight down at a river far below with no sign of the road or even the shoulder, while those on the other side watched giant, jagged boulders swing by inches away. After we got past them, it was a straight shot down the mountain to the town of War.

War had seen better days. Its main street consisted of some tired old stores, a bank, a couple of gas stations, and a crumbling hotel. During the 1920's, according to the history the War kids recounted from their parents, War was a wild, bawdy place of dance halls and gambling houses. Maybe that was why whenever a lady wore too much perfume my mother would say she smelled "like Sunday morning in War."

Big Creek High School sat on the outskirts of War beside the river that gave the district its name. It was a grimy three-story brick building with a carefully tended football field in front. On the other side of the football field was a train track. Our classes were often interrupted by the rumble of coal cars and the moans of steam locomotives going past. Sometimes it seemed as if they would never stop, train after endless train bound for the world that lay beyond us.

After getting to Big Creek, we usually had an hour to wait before classes, and Roy Lee, Sherman, O'Dell, and I spent the time together in the auditorium, trading homework and watching the girls parade up and down the aisle. That morning, I wanted to sit down with them and talk about algebra. I hadn't figured out the assigned problems to my satisfaction. But nobody else wanted to talk about algebra, not with *Sputnik* to chew over. "The Russians aren't smart enough to build a rocket," Roy Lee said. "They must've stole it from us." I didn't agree with him and said so. The Russians had built atomic and hydrogen bombs, and they had jet bombers that could reach the United States. So why couldn't they build something like *Sputnik* too?

"I wonder what it's like to be a Russian?" Sherman asked, aware that none of us had the slightest notion. Sherman was always wondering what it was like to live somewhere else other than West Virginia. I never gave it that much thought at all. I figured one place was like another, except, according to the television, if you lived in New York or Chicago or any big city, you had to be plenty tough.

Roy Lee said, "My daddy said the Russians ate their own babies in the war and it was a good thing the Germans attacked them. He said we should have joined the Germans and kicked their tails. Then we wouldn't be having so much trouble with them now."

O'Dell had been eyeing a senior cheerleader standing in the aisle. "I wonder if I crawled over there and kissed her feet if she'd pet me on the head?" he mused.

"Her boyfriend might," Sherman said as a huge football player stalked up and took her hand. Football players more or less had their pick of the girls at Big Creek.

I said, desperately, "Did anybody get the algebra?"

The other three just looked at me. "Did you get the English?" Roy Lee finally asked.

I had—a bunch of diagramed sentences. We traded, talking over the work as we busily copied. It wasn't exactly cheating, and it was the only way I was going to get any points at all in algebra class. Mr. Hartsfield, Big Creek's math teacher, never gave partial credit in a test. The work was either right or wrong. It seemed the more frustrated I got, the wronger I tended to be, in algebra or anything else.

Sputnik came up as a topic again later in the day during Mr. Mams's biology class. At the time, I was contemplating a long pickled worm stretched out in a square steel pan. To my everlasting delight, I had somehow managed to get Dorothy Plunk as my partner for the worm dissection. It was my opinion that Dorothy Plunk, a native of War, was the most beautiful girl in our class or, for that matter, at Big Creek High. She had a long shimmering ponytail and eyes the electric blue of my father's 1957 Buick. She also had a budding figure that made me feel as if I was going to explode. I had shyly managed to say hello to her in the hall a few times, but hadn't figured out any way to hold a real conversation with her. I couldn't even figure out what to say to her over the dead worm we were supposed to cut up together. The crackle of the intercom system intervened before I could come up with anything. The voice we heard was that of our school principal, Mr. R. L. Turner:

"As I'm sure you know by now," Mr. Turner said in his deliberate manner, "the Russians have launched a satellite into space. There have been many calls for the United States to do something in response. The Big Creek Student Council today has responded to, and I quote, the 'threat of *Sputnik*' by passing a resolution—I have it in my hand now—that dedicates the remainder of the school year to academic excellence. I approve the council's resolution. That is all."

Dorothy and I had been staring at the intercom. When we looked down, we were facing each other and our eyes locked. My heart did a little flip-flop. "Are you scared?" she asked me.

"Of the Russians?" I gulped, trying to breathe. The truth was, at that moment Dorothy scared me a lot more than a billion Russians, and I didn't know why.

She gave me a soft little smile, and my heart wobbled off its axis. I could smell her perfume even over the formaldehyde. "No, silly. Cutting open our worm."

Our worm! If it was *our* worm, couldn't it also be *our* hearts, *our* hands, *our* lips? "Not me!" I assured her and raised my scalpel, waiting for Mr. Mams to give us the go-ahead. When he did, I made a long cut down the length of the specimen. Dorothy took one look, grabbed her mouth, and lurched out the door, her ponytail flying. "What'd you do, Sonny?" Roy Lee chortled from the desk behind me. "Ask her for a date?"

I had never asked any girl out, much less the exalted Dorothy Plunk. I turned to Roy Lee and whispered, "Do you think she'd go out with me?"

Roy Lee wiggled his eyebrows, a leer on his face. "I got a car, and it's got a backseat. I'm your driver anytime you want."

Emily Sue Buckberry, who was Dorothy's best friend, stared at me, doubt written all over her round face. "She's got a boyfriend, Sonny," she said pointedly. "A couple of them. One's in college."

Roy Lee countered, "Aw, they're no competition. You don't know Sonny when he gets going. He's all action in the backseat."

My face flushed at Roy Lee's bragging. I'd never actually been with a girl in a backseat or anywhere else. The best I'd ever done was a kiss on a girl's front porch after a dance, and that was only with Teresa Anello in junior high school, just once. I turned back to the worm and made another cut and began to pin back the worm's flesh, taking meticulous notes. I thought to myself, Roy Lee just doesn't understand. Dorothy Plunk was no mere girl. Could he not see, as I, that Dorothy Plunk was God's perfection? She was to be worshiped, not handled. Happy in my daydream, I cut and wrote, wrote and cut. I was inspired. I was doing the work for Dorothy, my partner on this worm—and maybe more. Over the remains of a giant formaldehyde-soaked worm, I made up my mind to win her.

Roy Lee sneaked around my table and stared at my blissful expression. "Gawdalmighty," he complained. "You're in love."

Emily Sue came up on the other side. "I think you're right," she said. "This is serious."

"Heartbreak coming?" Roy Lee asked, as if from one professional in the love business to another.

"Undoubtedly," Emily Sue replied. "Sonny? What day is it, Sonny? Hello?"

I ignored them. A single name was the only lyric to the song in my brain. Over and over again it played: *Dorothy Plunk, Dorothy Plunk.*

THE Big Store steps was a favorite place for off-shift miners to lounge about, chew tobacco, and gossip. When a topic—especially one that happened outside Coalwood and also didn't involve mining or football—reached the steps, you knew it was important. *Sputnik* made it by midweek after its launch. I was going inside the store to buy a bottle of pop when I heard one of the miners on the steps say, "We ought to just shoot that damn Sputnikker down." There was a pause while the men all thoughtfully spat tobacco juice into their paper cups, and then one of them said, "Well, I'll tell you who we oughta shoot. Makes me madder'n fire"—he pronounced the word as if it rhymed with *tar*—"them damn people up in Charleston who's tryin' to cheat Big Creek out of the state champs. I'd like to warp them up side the head." This got even a louder affirmation from the assembly, followed by some truly hearty spitting. Only coal mining was more important in Coalwood than high-school football. *Sputnik*, and anything else, was going to always come in a distant third.

What made the miner "madder'n fire" was that Big Creek was on its way to an undefeated season, but according to the West Virginia High School Football Association, it was ineligible for the state championship because it played too many Virginia schools. On the mantrip cars into the mine, at the company stores, and even in church, this was a topic of endless discussion and debate. Big Creek kept winning, and the people in charge of high-school football up in Charleston kept saying it didn't matter—there was no way we were going to be state champs. It didn't take much of a genius to see there was some kind of trouble ahead. As it turned out, it was my dad who ended up causing the trouble.

BROTHER Jim was a fury on the football field. He played tackle on offense and linebacker on defense, and opposing quarterbacks ran from him like scared rabbits. He could hit like a locomotive and was a devastating blocker. At the time, a player as good as Jim was accorded nearly the same celebrity status across Big Creek district as Johnny Unitas and Bart Starr in the outside world. My father, utterly thrilled by Jim's gridiron prowess, was elected as president of the Big Creek Football Fathers' Association. I was watching television in the living room one night when Mom suggested to Dad, after he had spent some minutes on the mine phone (which we called the black phone) boasting about Jim to one of his foremen, that it might be a good thing if he bragged on me every once in a while. Even though he knew I was in the same room, Dad thought for a moment and then wondered aloud, quite honestly, "What about?"

I'm sure I didn't know either. I had no proclivity for football whatsoever. For one thing, I was terribly nearsighted. When I was in the third grade, Doc Lassiter came up to the school with an eye chart, and all of the children in my class were put in a line to read it. Our mothers, alerted by the school, were also there. I had most of the letters memorized by the time it was my turn, but Doc fooled me by putting up another chart. All I could see was a grayish blur. Doc gently told me to walk ahead until I could see the top letter. I walked forward until my nose nearly touched the wall. *"E!"* I announced proudly while Mom sobbed and the other mothers comforted her.

I tried out for the team at Coalwood Junior High for three straight years, but there was no way I was ever going to be anything more than a tackle dummy. "Sonny's small," Coach Tom Morgan told my Uncle Clarence at practice one day, "but he makes up for it by being slow." Everybody on the sidelines got a good laugh over that one. Quitting, however, never entered my mind. My mother would have dragged me right back to practice. It was one of her rules: *If you start something, you've got to finish it.*

When I went to Big Creek, Coach Merrill Gainer, the winningest coach in southern West Virginia history, took one look at me lost inside the practice gear and ordered me off his football field. I joined Big Creek's marching band as a drummer. Mom said she liked my uniform. Dad had no comment. Jim was mortified enough to complain about it at the supper table. While simultaneously chewing two huge spoonfuls of mashed potatoes, he explained the general lack of masculinity of boys who played in the band: "Boys don't play onna team gotta be chicken. Boys play inna band gotta be *real* chicken!" Jim worked for a little while more on the potatoes, swallowed, and then noted, "My brother's a sister."

"Well, my brother's an idiot," I responded reasonably and, to my way of thinking, objectively.

"If you two boys can't say anything nice at the table," Mom said with an utter lack of passion, "I'd just as soon you said nothing at all."

Jim's words had stung, but I shut up. I couldn't understand what all the interest in football was, anyway, and especially why the football boys were considered heroes. They were out on a field with a referee who made sure everybody followed all the rules, and the players wore pads on their shoulders and hips and thighs and knees, and helmets on their heads. What was heroic about lining up and following the rules and wearing a bunch of stuff that was going to keep you from getting hurt? I just never could understand it.

Dad remained silent at the table, but I noticed he and Jim exchanged a look of what I took to be agreement about the shame of me being in the band. I looked over at Mom for support, but she was looking through the window behind me. I supposed there were birds at her feeder. I thought to myself, I like the uniform and I like playing the snare drum. *And Dorothy Plunk's in the band too.* That last thought made me give Jim a smug look that confused him no end.

ALL that fall, the *Welch Daily News* and the *Bluefield Daily Telegraph* were filled with stories of our American scientists and engineers at Cape Canaveral in Florida, desperately working to catch up with the Russians. It was as if the science fiction I had read all my life were coming true. Gradually, I became fascinated by the whole thing. I read every article I could find about the men at the Cape and kept myself pinned to the television set for the latest on what they were doing. I began to hear about one particular rocket scientist named Dr. Wernher von Braun. His very name was exotic and exciting. I saw on television where Dr. von Braun had given an interview and he said, in a crisp German accent, that if he got the go-ahead he could put a satellite into orbit within thirty days. The newspapers said he'd have to wait, that the program called Vanguard would get the first chance. Vanguard was the United States's International Geophysical Year satellite program, and von Braun, since he worked for the Army, was somehow too tainted by that association to make the first American try for orbit. At night before I went to sleep, I thought about what Dr. von Braun might be doing at that very moment down at the Cape. I could just imagine him high on a gantry, lying on his back like Michelangelo, working with a wrench on the fuel lines of one of his rockets. I started to think about what an adventure it would be to work for him, helping him to build rockets and launching them into space. For all I knew, a man with that much conviction might even form an expedition into space, like Lewis and Clark. Either way, I wanted to be with him. I knew to do that I'd have to prepare myself in some way, get some skills of some kind or special knowledge about something. I was kind of vague on what it would be, but I could at least see I would need to be like the heroes in my books—brave and knowing more than the next man. I was starting to see myself past Coalwood. *Wernher von Braun. Dorothy Plunk.* My song now had two names in it.

When the papers printed that *Sputnik* was going to fly over southern West Virginia, I decided I had to see it for myself. I told my mother, and pretty soon the word spread, fence to fence, that I was going to look at *Sputnik* and anybody else who wanted to could join me in my backyard the evening it was scheduled to appear.

It didn't take much in Coalwood to create a gathering. On the appointed night, Mom joined me in the backyard, and then other women arrived and a few small children. Roy Lee, Sherman, and O'Dell were there too. The ladies clustered around Mom and she held court. Since Dad was who he was, she could always be counted on to know the latest on what the company was planning and which foreman was up and which was down. Watching her, I couldn't help but be proud at how pretty she was. Later in life, looking back on those days, I realized she was more than pretty.

Mom was beautiful. When she smiled it was like a hundred-watt bulb just got switched on. Her curly hair fell past her shoulders, she had big, hazel-green eyes, and her voice, when she wasn't using it to keep me and Jim straight, was soft and velvety. I don't think there was a miner in town who could get past the front gate when she was out in the yard in her shorts and halter tending her flowers. They'd stand there, tipping their helmets, grinning with their chewing tobacco–stained teeth. "Hidy, Elsie, them flares sure are lookin' good, that's for sure," they'd say. But I don't think they were looking at the flowers.

It got darker and the stars winked on, one by one. I sat on the back steps, turning every few seconds to check the clock on the kitchen wall. I was afraid maybe *Sputnik* wouldn't show up and even if it did, we'd miss it. The mountains that surrounded us allowed only a narrow sliver of sky to view. I had no idea how fast *Sputnik* would be, whether it would zip along or dawdle. I figured we'd have to be lucky to see it.

Dad came outside, looking for Mom. Something about seeing her out there in the backyard with the other women looking up at the stars vexed him. "Elsie? What in blue blazes are you looking at?"

"*Sputnik*, Homer."

"Over West Virginia?" His tone was incredulous.

"That's what Sonny read in the paper."

"President Eisenhower would never allow such a thing," he said emphatically.

"We'll see," Mom intoned, her favorite phrase.

"I'm going—"

"To the mine," my parents finished in a chorus.

Dad started to say something, but Mom raised her eyebrows at him and he seemed to think better of it. My father was a powerfully built man, standing just under six feet tall, but my mother could easily take his measure. He plopped on his hat and trudged off toward the tipple. He never looked up at the sky, not once.

Roy Lee sat down beside me. Before long, he was offering me unwanted advice on how to gain my beloved, Dorothy Plunk. "What you do, Sonny," he explained, putting his arm around my shoulders, "is take her to the movies. Something like *Frankenstein Meets the Wolf Man*. Then you kind of put your arm on the back of her chair like this, and then when things get scary and she's not paying attention to anything but the movie, you let your hand slide down over her shoulder until . . ." He pinched one of my nipples and I jumped. He laughed, holding his stomach and doubling over. I didn't think it was so funny.

Jim wandered outside and contemplated Roy Lee and me. He was eating a Moon Pie. "Idiots," he concluded. "Tenth-grade morons." Jim always had such a way with words. He squashed the entire pie in his mouth and chewed it contentedly. One of the neighbor girls down the street saw him and came over and stood as close to him as she dared. He smirked and rubbed his hand along the small of her back while she shivered in nervous delight. Roy Lee stared in abject admiration. "I don't care if they break every bone in my body, I *got* to go out for football next year."

"Look, look!" O'Dell suddenly cried, jumping up and down and pointing skyward. *"Sputnik!"*

Roy Lee sprang to his feet and yelled, "I see it too!" and then Sherman whooped and pointed. I stumbled off the steps and squinted in the general direction everybody was looking. All I could see were millions of stars. *"There,"* Mom said, taking my head and sighting my nose at a point in the sky.

Then I saw the bright little ball, moving majestically across the narrow star field between the ridgelines. I stared at it with no less rapt attention than if it had been God Himself in a golden chariot riding overhead. It soared with what seemed to me inexorable and dangerous purpose, as if there were no power in the universe that could stop it. All my life, everything important that had ever happened had always happened somewhere else. But *Sputnik* was right there in front of my eyes in my backyard in Coalwood, McDowell County, West Virginia, U.S.A. I couldn't believe it. I felt that if I stretched out enough, I could touch it. Then, in less than a minute, it was gone.

"Pretty thing," Mom said, summing up the general reaction of the backyard crowd. She and the other ladies went back to talking. It was a good hour before everybody else wandered off, but I remained behind, my face turned upward. I kept closing my mouth and it kept falling open again. I had never seen anything so marvelous in my life. I was still in the backyard when Dad came home. He opened the gate and saw me. "Aren't you out late?"

I didn't reply. I didn't want to break the spell *Sputnik* had cast over me.

Dad looked up at the sky with me. "Are you still looking for *Sputnik*?"

"Saw it," I said finally. I was still so overwhelmed I didn't even tag on a "sir."

Dad looked up with me for a little longer, but when I didn't elaborate he shook his head and went down into the basement. I soon heard the shower running and the sound of him scrubbing with brushes and Lava soap. He'd already showered at the mine, most likely, but Mom wouldn't let him in the house if he had a molecule of coal anywhere on him.

That night, in my room, I kept thinking about *Sputnik* until I couldn't think about it anymore and fell asleep, waking in the night to hear the men miners scuffling their boots and talking low as they went up the path to the tipple. I climbed up on my knees and looked through the window at their dark shapes walking alongside the road. The hoot-owl miners were the safety and rock-dust crew, assigned the task to spray heavy rock powder into the air to hold down the explosive coal dust. They also inspected the inside track, the support timbers, and the roof bolts. It was their job to make certain the mine was safe for the two coal-digging shifts. The way they looked in the moonlight, slogging in the dust, I could imagine them to be spacemen on the moon. The tipple, lit up by beacons, could have been a station there. I let my imagination wander, seeing the first explorers on the moon as they worked their way back to their station after a day of walking among craters and plains. I guessed it would be Wernher von Braun up there, leading his select crew. The men crossed the tracks and I saw the glint of their lunch buckets in the tipple light, and I came slowly back to reality. They weren't explorers on the moon, just Coalwood miners going to work. And I wasn't on von Braun's team. I was a boy in Coalwood, West Virginia. All of a sudden, that wasn't good enough.

———

ON November 3, the Russians struck again, launching *Sputnik II.* This one had a dog in it—Laika was her name—and by her picture in the paper, she looked a little like Poteet. I went out into the yard and called Poteet over and picked her up. She wasn't a big dog, but she felt pretty heavy. Mom saw me and came outside. "What are you doing to that dog?"

"I just wondered how big a rocket it would take to put her into orbit."

"If she don't stop peeing on my rosebushes, she's going into orbit, won't need any rocket," Mom said.

Poteet whined and ducked her head in my armpit. She might not have known every word, but she knew very well what Mom was saying. As soon as Mom went back inside, I put Poteet down and she went over and sat by one of the rosebushes. I didn't watch to see what she did after that.

My dad got two magazines in the mail every week, *Newsweek* and *Life.* When they came, he read them from cover to cover and then I got them next. In a November issue of *Life,* I found, to my great interest, drawings of the internal mechanisms of a variety of different kinds of rockets. I studied them carefully, and then I remembered reading how Wernher von Braun had built rockets when he was a youngster. An inspiration came to me. At the supper table that night, I put down my fork and announced that I was going to build a rocket. Dad, musing into his glass of corn bread and milk, said nothing. He was probably working through some ventilation problem, and I doubt if he even heard me. Jim snickered. He probably thought it was a sister thing to do. Mom stared at me for a long while and then said, "Well, don't blow yourself up."

I gathered Roy Lee, O'Dell, and Sherman in my room. My mom's pet squirrel, Chipper, was hanging upside down on the curtains, watching us. Chipper had the run of the house and loved to join a gathering. "We're going to build a rocket," I said as the little rodent launched himself at my shoulder. He landed and snuggled up against my ear. I petted him absently.

The other boys looked at one another and shrugged. "Where will we launch it?" was all that Roy Lee wanted to know. Chipper wiggled his nose in Roy Lee's direction and then hopped off my shoulder to the bed and then to the floor. The sneak attack was Chipper's favorite game.

"The fence by the rosebushes," I said. My house was narrowly fitted between two mountains and a creek, but there was a small clearing behind Mom's rose garden.

"We'll need a countdown," Sherman stated flatly.

"Well of course we *have* to have a countdown," O'Dell argued, even though no one was arguing with him. "But what will we make our rocket out of? I can get stuff if you tell me what we need." O'Dell's father—Red—was the town garbageman. On weekends, O'Dell and his brothers helped out on the truck and saw just about every kind of stuff there was in Coalwood, one time or another.

Sherman was always a practical boy with an orderly mind. "Do we know *how* to build a rocket?" he wondered.

I showed them the *Life* magazine. "All you have to do is put fuel in a tube and a hole at the bottom of it."

"What kind of fuel?"

I had already given the matter some thought. "I've got twelve cherry bombs left over from the Fourth of July," I said. "I've been saving them for New Year's. We'll use the powder out of them."

Satisfied, Sherman nodded. "Okay, that ought to do it. We'll start the countdown at ten."

"How high will it fly?" O'Dell wondered.

"High," I guessed.

We all sat around in a little circle and looked at one another. I didn't have to spell it out. It was an important moment and we knew it. We boys in Coalwood were joining the space race. "All right, let's do it," Roy Lee said just as Chipper landed on his D.A. Roy Lee leapt to his feet and flailed ineffectually at his attacker. Chipper giggled and then jumped for the curtain.

"Chipper! Bad squirrel!" I yelled, but he just closed his beady eyes and vibrated in undisguised delight.

Roy Lee rolled up the *Life* magazine, but before he could raise his arm, Chipper was gone in a flash, halfway down the stairs toward the safety of Mom in the kitchen. "I can't wait for squirrel season," Roy Lee muttered.

I appointed myself chief rocket designer. O'Dell provided me with a small discarded plastic flashlight to use as the body of the rocket. I emptied its batteries and then punched a hole in its base with a nail. I cracked open my cherry bombs and poured the powder from them into the flashlight and then wrapped it all up in electrical tape. I took one of the cherry-bomb fuses left over and stuck it in the hole and then glued the entire apparatus inside the fuselage of a dewinged plastic model airplane—I recall it was an F-100 Super Sabre. Since Sherman couldn't run very fast—and also because it was his idea—he was placed in charge of the countdown, a position that allowed him to stand back. Roy Lee was to bring the matches. O'Dell was to strike the match and hand it to me. I would light the fuse and make a run for it. Everybody had something to do.

When night came, we balanced our rocket, looking wicked and sleek, on top of my mother's rose-garden fence. The fence was a source of some pride and satisfaction to her. It had taken six months of her reminding Dad before he finally sent Mr. McDuff down from the mine to build it. The night was cold and clear—all the better, we thought, for us to track our rocket as it streaked across the dark, starry sky. We waited until some coal cars rumbled past, and then I lit the fuse and ran back to the grass at the edge of the rosebushes. O'Dell smacked his hand over his mouth to smother his excited giggle.

Sparkles of fire dribbled out of the fuse. Sherman was counting backward from ten. We waited expectantly, and then Sherman reached zero and yelled, *"Blast off!"* just as the cherry-bomb powder detonated.

There was an eyewitness, a miner waiting for a ride at the gas station across the street. For the edification of the fence gossipers, he would later describe what he had seen. There was, he reported, a huge flash in the Hickam's yard and a sound like God Himself had clapped His hands. Then an arc of fire lifted up and up into the darkness, turning and cartwheeling and spewing bright sparks. The way the man told it, our rocket was a beautiful and glorious sight, and I guess he was right, as far as it went. The only problem was, it wasn't our rocket that streaked into that dark, cold, clear, and starry night.

It was my mother's rose-garden fence.

3

M O M

WOODEN SPLINTERS WHISTLED past my ears. Big chunks of
the fence arced into the sky. Burning debris fell with a clatter. A
thunderous echo rumbled back from the surrounding hollows.
Dogs up and down the valley barked and house lights came on,
one by one. People came out and huddled on their front porches.
Later, I would hear that a lot of them were wondering if the mine
had blown up or maybe the Russians had attacked. At that moment, I wasn't thinking about anything except a big orange circle
that seemed to be hovering in front of my eyes. When I regained
some sensibility and my vision started to come back, the circle
diminished and I started to look around. All the other boys were
sitting in the grass, holding their ears. With relief, I noted that it
didn't look as if any of them had suffered any serious damage. Roy
Lee's D.A. needed work though, and O'Dell's eyes were as wide as
the barn owls that nested on the tipple. Sherman's glasses were
nearly sideways on his face. The dogs had retreated to the farthest
corner of the yard. They were crawling on their bellies back toward
us when Mom came out on the back porch and peered into the
darkness. "Sonny?" she called. Then I think she saw the burning
fence. "Oh, my good Lord!"

Dad, holding his newspaper, came out beside her. "What happened, Elsie?"

At my father's appearance, the other boys suddenly jumped up
and ran off. I guess he had such a fierce reputation at the mine they

didn't want any part of his wrath. I fleetingly caught a glimpse of Roy Lee leaping over the still-standing part of the fence, clearing it by a good yard. The others went through the gap we'd just blown out. I could see them clearly because the standing part of the fence was on fire. I thought to myself, I ought to follow them, maybe take up residence in the woods for a year or two. But I was caught. Running would just put off the inevitable. I answered Mom with a croak, my mouth not working quite right yet. She replied, "Sonny Hickam. You get over here!" Rubbing my ears in an attempt to stop them from ringing, I lurched over to the back porch and waited expectantly for one of my parents to come down off it and kill me.

"Elsie, do you have any idea what's going on here?" Dad asked.

Mom, bless her, had figured it all out. "Sonny asked us if he could build a rocket, Homer," she replied, as if she were amazed he had not perceived the perfectly obvious.

Dad puzzled over her statement. "Sonny built a rocket? He doesn't even know how to put the sprocket chain back on his bike when it slips off."

"We'll see," Mom sniffed. "Sonny, what happened to the other boys?"

I had learned that sometimes when I was in trouble with Mom, the best thing to do was to adopt the complete-idiot strategy. "Other boys?" I asked, most sincerely. Even under the greatest duress, my capability to dissemble was scarcely diminished. Once, when I had used Mom's best and only wheelbarrow as a kind of summertime sled to go careening down a gully on Substation Mountain, and then misplaced the legs I had removed and the screws that bolted them on, and then dented the barrow almost beyond recognition on a boulder that popped up in my way, and flattened the tire of the wheel, what I'd said then when I came home with the remnants of the thing was that I'd spotted some great flower dirt up in the mountains and would've brought Mom some home with me "if this blame ol' 'wheelbare' hadn't fallen apart!" Mom wasn't fooled, but she got to laughing too hard to swat me at full power. Whatever it took, sometimes, is what I did.

"Elsie, I don't care about any other boys," Dad told her. "Just

take care of this one before he embarrasses me all over Coalwood."

Mom laughed—a short, bitter bark. "Oh, my, yes. Heaven for-
bid you be embarrassed! Why, the next thing you know, the men
would stop shoveling coal for you!"

He stared at her. "They don't shovel. They haven't shoveled in
twenty years. They use machines."

"Isn't *that* interesting!"

I recognized that Mom and Dad were about to go off onto one
of their standard quarrels and eased back into the darkness of the
yard and stood with the dogs. Dandy nuzzled my hand and Poteet
leaned against my legs. I could feel her trembling, or maybe it was
me. Dad gave Mom one of his standard speeches about how the
mine provided for her and us boys, and she said back her usual
piece about how the mine was just a big, dirty death trap. When
Dad went back inside, shaking his head, Mrs. Sharitz next door
called softly to Mom and she went over and leaned on the fence. I
couldn't hear what they were saying, but I could guess. I could see
Mrs. Todd waiting patiently at the next fence beyond. Mrs. Sharitz
would cross her yard with the news from Mom and pass it on to
Mrs. Todd and so on down the fence line. I knew within an hour
all of Coalwood would know about my semi-sort of rocket and
how I'd roped the other boys into more of my foolishness, and
everybody in town would have a good laugh at my expense. When
Mom signed off with Mrs. Sharitz, she walked over and stood
beside me. She looked at the smoldering ruin of her fence and
sighed deeply. I braced myself. Now that we were alone, she
was free to deliver her scorn with both barrels. "Didn't I tell you
not to blow yourself up?" she asked in a surprisingly soft
voice.

Just then, I heard the black phone ring and saw Dad through
the living-room window as he ran to answer it. I hoped it wasn't
anybody complaining about the noise. Mom looked at the window
and then up the road to the tipple. I knew the best thing I could
do was to stay quiet while she was chewing things over in her
mind. After a while, she pointed at the back porch and said, "Go
sit on the steps. We need to talk, you and me."

"I know what I did was wrong, Mom," I said in a bid to preempt whatever she had in mind.

"Homer Hadley Hickam, Junior. It wasn't wrong. It was stupid. *I said go sit!*"

I did as I was told with the enthusiasm of a prisoner going to his own beheading. Dandy crawled up beside me, whimpered briefly, and laid his head on my feet. Poteet was off chasing bats. I watched her launch herself into the air, do a double twist, and come down running, a big grin on her black muzzle.

I thought to myself, I'm really in for it this time. Mom was a master at delivering creative punishment. Once, after Sunday school, and in my usual rush to get outside and play, I wore my church shoes in the creek to hunt crawl-dads with Roy Lee. When Mom cast an eye on my soggy Buster Browns, she said, "I swear, Sonny, if your head gets any emptier, it's going to float off your head like a balloon." For punishment, she dictated that the next week I had to go to church in my stocking feet. It didn't take long before everybody in town got wind of what I was going to have to do. I didn't disappoint, walking down the church aisle in my socks while everybody nudged their neighbor and snickered. The thing was, though, I had picked out the socks, and my big toe poked through a hole in one of them. Mom was mortified. Even the preacher couldn't keep a straight face.

Mom stood before me and crossed her arms and stuck her chin out. Dad said she looked just like a Lavender when she did that, and it usually always meant trouble. "Sonny, do you think you could build a real rocket?"

She so startled me by her question that I forgot my usual coyness. "No, ma'am," I said, straight up. "I don't know how."

She rolled her eyes. "I know you don't know how. I'm asking you if you put your mind to it, could you do it?"

I searched for her trap to make me do something I didn't want to do. I was sure it was there. It was just a matter of finding it. I thought I'd better say something. "Well, I guess I could—"

Mom stopped me. She knew I was just going to ramble. "Sonny," she sighed, "you're a sweet kid. I love you. But, doggone it, you've just been drifting along like you were on a cloud your whole life, making up games and leading Roy Lee and Sherman and O'Dell off on all your wild schemes. I'm thinking maybe it's past time you straightened up a bit."

When a Coalwood mother told her son maybe he needed to "straighten up a bit," it was usually in a direction he didn't necessarily want to go. I started to squirm. She was about to make it ten times worse. "I was worrying about you the other night to your dad," she said. "I was just kind of wondering out loud what you were going to do with yourself when you grew up. He said for me not to worry, he'd find you a job on the outside up at the mine. You know what that means, Sonny? You'd be some kind of clerk working for your dad, sitting at a typewriter pecking out forms, or writing in a ledger about how many tons got loaded in a day. That's the best your dad thinks you can do."

A question just seemed to jump out of my mouth. It surprised even me. I guess I'd been wondering about it for a long time and didn't know it. "Why doesn't Dad like me?" I asked her.

Mom looked as if I had slapped her in the face. She was quiet for a moment, obviously chewing over my question. "It's not that he doesn't like you," she said at length. "It's just with the mine and all, he's never had much time to think about you one way or the other."

If that was supposed to make me feel any better, it didn't work. I knew Dad thought about Jim all the time, was always telling people what a great football player my brother was, and how he was going to tear up the world in football when he went to college.

Mom sat down beside me and put her arm around my shoulders. I twitched at her unfamiliar touch. It had been a long time since she'd hugged me. We just didn't do much of that kind of thing in our family. "You've got to get out of Coalwood, Sonny," Mom said. "Jimmie will go. Football will get him out. I'd like to see him a doctor, or a dentist, something like that. But football will get him out of Coalwood, and then he can go and be anything he wants to be."

She clutched my shoulder, pulling me hard against her side. For a little bit, I would've put my head on her shoulder, but I sensed that would be going too far. "It's not going to be so easy for you," she said. "You and me, we've got to figure out some way to make your dad change his mind about you, see his way clear to send you to college. I've been saving money right along and probably have enough for you to go, but your dad would have a hissy right now if I said that's what I was going to do, say it's a waste of good money. He's got it in his head you're going to stay around here, have some little job at his mine."

"I'd like to go to college, sure—" I began.

"Well, you'd better!" she snapped, cutting me off. She dropped her arm off my shoulder. I felt suddenly chilled without it. "Coalwood's going to die," she announced, "deader than a hammer."

"Ma'am?" She had lost me with that one.

She stood up. I saw that her eyes were glistening. By her nature, she wasn't a crier. In an instant, she brought herself back under complete control. "You can't count on the mine being here when you graduate from high school, Sonny. You can't even count on this town being here. Pay attention, will you? Look at the kids at Big Creek from Berwind, Bartley, Cucumber. . . . Their fathers are out of work, and those towns are just falling down around them. It's the economy and it's the easy coal playing out and it's . . . I don't know what all it is, but I've got sense enough to know it's just a matter of time before the same thing happens here in Coalwood too. You need to do everything you can to get out of here, starting right now."

I didn't know what to say. I just stared at her. She sighed. "To get out of here, you've got to show your dad you're smarter than he thinks. I believe you can build a rocket. He doesn't. I want you to show him I'm right and he's wrong. Is that too much to ask?"

Before I could reply, she sighed deeply once more, glanced over at her fence (the fire had burned itself out), and then stomped past me and went inside. I eased my foot out from under Dandy's head so he wouldn't be disturbed and came off the steps and stood alone in the deep blackness of the backyard, the old mountains looming over me. I tried to think, to catch up with all she had said. Dandy got up and sidled over to me and licked my hand. He was a good old dog. Poteet had stopped chasing bats and was asleep under the apple tree.

When I went inside, Dad was still on the black phone. He said nearly everything over the company phone in exclamations. "Get Number Four back on line, and I mean now!" Number Four was undoubtedly one of the huge ventilators on the surface that forced air through the mine. Whoever was on the other end apparently wasn't giving him the response he wanted. "I'm leaving the house right now, and it better be going by the time I get there!" He slammed the receiver down. I watched him throw open the hall closet and snatch his jacket and hat. He rushed past me without a glance, just as if I didn't exist, and went out the back door. I heard the gate unlatch and close, and he was gone into the night.

I went up the stairs and found Mom waiting for me in the hall. She wasn't through with me yet. "Has anything I've said tonight made any sense to you at all?"

I guess I looked blank. "Well . . ." I began.

"Oh, God, Sonny," she groaned in exasperation. She touched me on the nose with her finger. "I-am-counting-on-you," she said, tapping my nose with each word. "Show him you can do some-thing! Build a rocket!" Then she looked at me in a significant way and went inside her bedroom.

It was past midnight when Dad returned. I had just dozed off after a round of thinking about all that Mom had said. I heard him creep up the stairs and then I started thinking all over again. I carefully lifted Daisy Mae, my little calico cat, out of the crook of my arm and placed her at the foot of the bed and got up and opened my window. The tipple loomed before me like a giant black spider. According to Mom, Dad thought all I was good for was working there as a clerk. A gasp of steam erupted from an air vent beside the tipple, and I followed the cloud skyward, watched the water droplets disperse. A big golden moon hovered overhead, and the vapor formed a misty circle around it. Sparkling stars flowed down the narrow river of sky the mountains allowed. I looked at the tiny pricks of light so far away. I didn't know one star from the other, didn't know much of anything about the reality of space. I knew less than nothing about rockets too. I suddenly felt as stupid as Dad apparently thought I was. Mom had said for me to build a rocket, show him what I could do. I had already been thinking about learning enough to go to work for Wernher von Braun. Her Elsie Hickam Scholarship, if approved by Dad, would fit right in with that.

Then I remembered what Mom had said about Coalwood dying. That was the hardest thing to understand of all the things she had told me. All around me, Coalwood was always busily playing its industrial symphony of rumbling coal cars, spouting locomotives, the tromping of the miners going to and from the mine. How could that ever end?

The black phone interrupted my thoughts. Dad had probably just let his head touch his pillow when it rang. I heard his muffled voice as he answered it and then a string of what I was certain were curses. Within seconds, his door banged open and I heard him thumping down the stairs almost as if someone was chasing him. At the bottom of the stairs, he started to cough, a racking, deep, wet hack. He'd lately been complaining a lot about his allergies, even though in the fall you'd think there wouldn't be much pollen in the air. I'd often awoke to hear him coughing at night, but I'd never heard it so bad before. I watched him out of my bedroom window a few minutes later as he walked quickly toward the mine, his head down and a bandanna to his face. He stopped once and bent nearly double, a great spasm rocking him. Those allergies were really getting to him, I thought. He straightened and hurried on. As he neared the track, a long line of loaded coal cars trundled out of his way as if in recognition of his approach. As soon as he hopped the rails and disappeared up the path, the cars moved back to block my view. Mom's bedroom was beside mine, and I heard her pull her window shade down. She'd been watching him too.

4

THE FOOTBALL FATHERS

FOR THE NEXT week, the destruction of Mom's rose-garden fence by my rocket dominated conversation in Coalwood. Mr. McDuff came down from the mine to restore the fence and reported it had been reduced to splinters. "Maybe Elsie better get my Mister to build her next fence out of steel," Mrs. McDuff said to a friend at the Big Store. Soon ladies in their backyards were repeating that remark, fence by fence, from one end of the valley to the other. On the way to the tipple or in the man-trips, on the main line, in the gob (the mix of rock and coal dust that lay in the old part of the mine), and even at the face, the miners were talking about the great blast.

"You little sisters are idiot morons," Buck Trant, the big, ugly fullback, announced from the back bench of the morning school bus. He laughed at himself, thinking his observation brilliant. The other players joined in. *"Little idiot moron sisters!"*

Buck added, after a moment of concentration, "You sisters couldn't blow your noses without your mamas!"

Roy Lee, Sherman, and O'Dell bowed their heads in impotent rage. Not me. Buck Trant was too easy. Not only was he a dimwit, he was vulnerable. "At least we know where our mamas are," I shot back. Buck's mother had run off with a vacuum-cleaner salesman a few years before. Just as soon as I'd said my mean thing, I regretted it, but it was too late. Outraged, Buck jumped to his feet, but when Jack stomped on the brakes, he went tumbling. We were halfway up Coalwood Mountain. Without a word, Jack pulled the bus off the road, turned in his seat, and pointed at me. "Out!" he ordered. He looked at Buck. "You too, Buck!"

"Me?" Buck whined. "What did I do? Sonny started it. He's always starting stuff, you know that."

Jack didn't take guff off anybody on his school bus, even big overgrown football players. "Don't make me have to kick you out the door, son," he growled.

Buck looked for support from the other football boys, but they all had their heads down. He walked meekly down the aisle and got out, standing forlornly in the dirt. I followed him and we stood beside each other while Jack slammed the door shut. Before the bus rounded the curve, Buck was after me. I threw down my books, ducked his bear hug, and scampered up the mountainside and disappeared into the woods. "I'm going to murder you, you little four-eyed freak," he yelled after me.

"You and what army?" I challenged him from deep within a thicket of rhododendron. Buck huffed around along the road, but didn't come after me, probably because he was wearing blue suede shoes and didn't want to get them dirty. After a while, a car came along and Buck stuck his thumb out and climbed in. I came down and did the same, hitching to Big Creek, just making it in time for the first class. I avoided Buck all day, which was not easy since his locker was beside mine. Roy Lee and the other boys caught me at lunch. "We aren't going to build another rocket," Roy Lee said.

"Fine," I replied. I was already mad at him and the others for not backing me up on the bus. "I'll build one by myself!" I said it with such certainty, it surprised even me. Whether I liked it or not, I was committed to do it.

"Have at it," Roy Lee muttered, and he and O'Dell and Sherman walked away. I knew I'd really messed up. I needed their help. I had to build a rocket and I didn't have a clue where to start.

———

THAT night, while I was puzzling over my algebra, Jim stuck his head in my room. "I just want you to know how really great it is to have a brother who's a complete moron."

"Don't worry about it," I replied lamely.

"Everybody's laughing at the family because of you."

"Just go away," I growled. "I'm busy."

"Doing what?" he chided. "Trying to decide what dress to wear?"

Jim ducked when I threw my pencil at him and then pulled my door shut. Unbidden, a little bubble of brotherly jealousy gurgled up inside me. Who cared what Jim thought about anything? He didn't even have to think. Dad would take care of everything for him, see that he got everything he wanted. Jim thought I was some kind of a sister. Well, at least I didn't go around wearing pink shirts and a peroxide curl in my hair!

My first rocket had caused me to be harassed on the school bus, at school, and now in my own room. There was more to come. The following Saturday, when I went to the Big Store to buy a bottle of pop, I ran afoul of Pooky Suggs.

Pooky Suggs's history was common knowledge in Coalwood. His father had been crushed by a slate fall about a dozen years back on a section where Dad was the foreman. To stay in Coalwood, Pooky had quit the sixth grade and gone into the mine. To anybody who'd listen, Pooky was forever complaining about having to quit school to go to work, blaming it all on Dad for getting his daddy killed. He didn't get much sympathy. It had been his daddy's fault, after all, that he had gone under an unsupported part of the roof to urinate, and anyway, Pooky had already been in the sixth grade for five years when he quit. Nobody in town much thought he'd ever have reached the seventh. Still, for as long as I could remember, I'd heard Pooky's name around the house, Dad telling Mom about something that Pooky had done that was stupid, or that he'd caught him idling back in the gob again, and Mom telling Dad back he should just fire Pooky and get it done with. For some reason, Dad had never taken her up on it. Maybe he felt a little guilty about Pooky's father, I don't know, but he seemed to tolerate Pooky more than he did other complainers and idlers.

I avoided Pooky whenever I could, but hadn't noticed him among the men gossiping on the Big Store steps. "Well, looky what we got here—Homer's little rocket boy," he said nastily. "Heard the damn thang blew up. Did your daddy help you build it?"

The men sitting on the steps turned to look at me. They were all holding paper cups for their chewing-tobacco spit. "You gonna build another one?" asked Tom Tickle, one of the single miners who lived in the Club House.

Tom was friendly. "Yes, sir, I am," I said.

"Well, attaboy!" the step group chorused.

"Shee-it. All he can do is build a bomb," Pooky said.

"Well, it was a damn good bomb!" Tom laughed. Pooky stood up and kicked his way through the assembly. If he had hoped to heap scorn on me, it hadn't worked. He shoved his helmet back on his head and leaned into me, his breath mostly alcohol fumes. "You Hickams think you're so hot, but you ain't no better'n me or nobody else in this town."

"Sonny didn't say no different, Pooky," Tom said. "Whyn't you go sleep it off afore you get into trouble?"

Pooky turned, rocking unsteadily in his hard-toe boots. His face was all angles, with a sharp, pointed nose and a triangular chin covered with stubble. Despite the easy availability of Dr. Hale, the company dentist, his teeth were yellow and cracked. His voice was a whine that sounded like an untuned fiddle. "We need to go on strike, I'm tellin' ya. That bastard Homer's gonna work us all to death!"

"I don't believe work's ever going to kill you, Pook." Tom grinned, and the step miners erupted in laughter.

"All y'all can just go to hell!" Pooky muttered. He probably meant it to sound tough, but it came out sort of pitiful. I couldn't help but feel a little sorry for him. He gave me another dirty look. "Your daddy killed my daddy," he said. "I ain't never gonna forget that!"

Tom stood up and tugged Pooky away from me, turning him around and pointing him across the street. "You better get on home, Pook."

I took the opportunity to slip through the men to go inside the Big Store. I got my bottle of pop, then leaned on the counter and drank it slowly, watching through the glass doors what was going on outside. Pooky and Tom looked like they were dancing, with Pooky trying to come inside the Big Store and Tom turning him back around. To my relief, Tom finally won and Pooky staggered off. Soon afterward, all the men got up, their gossiping done. When the steps were clear, I ran outside and grabbed my bike and ped-aled toward home. Near the Coalwood School, I went past a line of miners making their way to the tipple. With big grins on their faces, they all yelled "Rocket boy!" as I flashed by. What had I gotten myself into? I'd told too many people I was going to build another rocket, and now I had to do it. But how? What was the blamed secret that made a rocket fly?

THE final regular season football game ended with Big Creek winning big over Tazewell High School, just across the Virginia border. Jim sent two quarterbacks to the sidelines on stretchers and intercepted a pass and ran it back for a touchdown. With that victory, the team had won all its games. Then the state high-school athletic association did exactly what it said it was going to do and ruled that Coach Gainer's boys were not eligible to play in the state-championship game. Although it was no surprise, there was still an instant uproar all over the district. The Football Fathers were besieged with demands from fans and the football team to do something. Jim asked Dad every night at supper for a week after the last game what he was going to do. Dad kept saying he was looking into it. Finally, one night at supper he said he was going to go see a lawyer in Welch.

Mom put down her fork and stared at Dad in disbelief. "Homer, I don't think that's wise."

Dad shoveled in a spoonful of beans and corn bread. "Elsie, I know what I'm doing," he replied nonchalantly. He didn't look at her.

Mom settled into a deep frown. "No, you don't. The Charleston muckety-mucks don't want us to play and they're not going to let us play. No lawyer's going to change that. You're just asking for trouble."

"Mom, Dad's got to do something!" Jim begged. "We deserve to play!"

"I know you do, Jimmie," Mom replied softly. "But sometimes we don't get our way even when we deserve it. That's true for everybody, even you. I know that amazes you, but that's the way it is."

Jim's face went dark and he shoved his chair back from the table. "I want to be excused," he said sullenly.

Dad held up his right hand to his face, as if to shield it from Mom's riveting gaze. "Jim, it's okay," he said reassuringly. "I'm going to take care of it."

"Homer—" Mom said in her warning tone.

"Elsie—" Dad said back in his don't-mess-with-me tone.

Jim stood up. "Somebody better do something!" he wailed.

I made my move. The football boys, even my brother, were so easy. "You could move to Charleston and play up there," I suggested virtuously.

Jim turned on me, his fists tightly balled. "You're dead, Sonny."

"Jim, go to your room," Mom ordered. She waited until Jim stalked off and then gave me a threatening look before turning back to Dad. "Homer, just let it go," she said.

Dad rolled his head on his neck. I could hear it creak. Ducking his head in the mine all day probably didn't do it a lot of good. "This is none of your affair, Elsie," he said.

"Just stop and think, that's all I'm asking."

"The Football Fathers—"

"If you put the brains of every one of the Football Fathers in my coffee cup, they wouldn't fill it up. You've got to think for all of them."

"We've made up our minds, Elsie. We're going to Welch."

Mom knew the Bible pretty much by heart, and she was quite capable of using it on Dad like a club. "And if the blind leadeth the blind, both shall fall into the ditch," she told him, making her argument unassailable because it was clearly on the side of the Lord.

After an evident moment of confusion, Dad replied, "Thank you for your vote of confidence, Reverend Lavender." Then the black phone rang, ever the convenient way to end discussion in our house. Dad yelled at whomever was on the other end and then headed out the door, throwing on his coat and hat as he went. I had no doubt he was grateful for the interruption. He didn't come home until past midnight. Sometimes when he did that, I wondered if he wasn't just sitting around his office, checking his watch, until Mom went to bed.

A week later, as Mom had feared, the Football Fathers had themselves a proud little suit. The championship game was only a week away, so it was necessary to press the court to act fast. Three days after it was filed, a state judge in Bluefield took one look at the case and threw it out of court on a technicality. There was no precedent, he wrote, for a private organization to sue a state entity in his court. The championship game was held as scheduled in Charleston, and the season was officially over. Jim was so mad he locked himself in his room all day, except to come downstairs to eat and watch television and talk to some girls on the home telephone. I kept out of his way, retreating to a chair in the living room to read Dad's latest *Newsweek*.

"I'm glad it's over," Mom said, watching Jim stomp morosely up and down the steps.

"We're going to appeal," Dad said from his easy chair. He was reading the paper. "We're going to go over that judge's head."

"But the game's been played!"

"It's the principle of the thing," Dad replied.

Mom went into the living room and stood over him. "What principle? It's high-school football!"

Dad turned the page, as if he had just finished an article. I noticed, from my position, that he had turned to the comic page, which he never read. When Mom kept looking at him, he finally said, his eyes still firmly planted on the funnies, "This is man's work, Elsie."

"Maybe so, Homer," Mom replied, "but this woman is telling you it's going to lead to disaster."

"We'll see," he said, stealing her phrase.

WINTER came to West Virginia late that year. It was a splendid fall; the leaves kept their bright burnt color well into November, and the sky turned a pale but pretty blue, like a robin's egg. Just before Thanksgiving, the first of the cold fronts from Canada finally reached us, and the trees abruptly dropped their leaves and turned black and skeletal. Winter storm clouds scudded in, got snagged on our hills, and stayed. Everything just seemed to turn black, brown, and gray after that.

Coalwood had its routine for the beginning of winter, just as it did for every season. Mrs. Eleanor Marie Dantzler, the wife of Mr. Devotee Dantzler, the company-store manager, started to plan her winter piano recital, an annual social event. Company coal trucks made the rounds of the houses, stocking the coal boxes. The Coalwood Women's Club built a float for the Veteran's Day parade in Welch. In 1957, Jim and the other football boys wore Marine uniforms and pretended to be raising the flag on Iwo Jima. A lot of veterans were seen sobbing on the Welch streets as it went by. Just behind the Coalwood float, the Big Creek band marched, with me proudly playing the snare drum, one of five drummers in a line. Standing with Mom on the curb, Dad clapped and cheered the Coalwood float as it went past. His eyes were on Jim the whole time. Before I got there, he turned to talk to somebody behind him and didn't see me as I paraded past. "Attaboy, Sonny," I heard Mom call above the *rat-a-tat-tat* of my drum.

THE union leader in Coalwood was a man named Mr. John Dubonnet, a classmate of my parents at Gary High School. During World War II, many Coalwood miners, including my father, were exempted from service because of the need for coal in the war effort. Mr. Dubonnet could have probably also stayed in West Virginia, but instead joined the Army. While he was landing on Normandy, my dad was opening up a new part of the mine, an incredibly rich vein of "high" coal, so-called because it was so thick a man could stand straight up in the tunnel left after its removal. By war's end, the Coalwood mine was a lucrative little operation, envied across the county. That was when the UMWA finally turned its attention to Mr. Carter's mine. The labor peace that Coalwood had enjoyed for more than fifty years came to an abrupt end. When Mr. Carter resisted the union's attempts to organize, the union ordered a strike. In retaliation, Mr. Carter instituted a lockout, closing the mine to everybody. There was some pushing and brawling around the tipple and rumors of gunfire up the hollows. To calm things down, President Truman sent in the United States Navy to reopen the mine. After six months of military occupation in Coalwood, Mr. Carter was forced to sign a contract with the union and, soon afterward, sold Coalwood in disgust. The Captain, and my father, stayed behind.

In the decade that followed, an edgy peace between labor and management settled on our town, broken only by intermittent strikes, usually quickly settled. The Coalwood operation became even richer. When the Captain retired, my father, at the Captain's insistence, took over his position of mine superintendent. Since he was only a high-school graduate, many people in Coalwood—and in the union and the steel company that owned us too—thought my father was not qualified. Dad set out to prove them wrong by the sheer volume of his work and application of every particle of his energy and intelligence. He also continued to carry the Captain's vision of the town long after nearly everyone else had forgotten it.

By 1957, most of the old union leaders had followed the Captain into retirement, and a new crop was eager to show their worth to the rank and file. Mr. Dubonnet was one of them, quickly rising to lead the Coalwood UMWA local. Although nobody noticed it until it was too late, having Mr. Dubonnet and my dad on opposite sides was a prescription for conflict.

As Mom had predicted, one day in early winter Dad stood outside the tipple and called out the names of men to be cut off. A national recession was under way, steel orders were reduced, and Coalwood was producing more coal than the steel company needed. Twenty-five men were cut off from the company. The phrase was apt. Not only were the men separated from their work, they were cut off from their homes, credit at the company stores, and identification as a Coalwood citizen. The miners cut off were required to leave their houses within two weeks. A few of them surreptitiously moved up past Snakeroot and built shacks along the fringe of the woods, hoping someday to be rehired. My father was later ordered by the steel company to bulldoze them out, but he never did. The church put together baskets of food at Thanksgiving and Christmas for these families. For the first time I could remember, I heard little children were showing up in Coalwood classrooms needing clothing and food.

After the cutoff, the union local, not certain what else to do about it, threatened to strike. Mr. Dubonnet showed up on our doorstep one evening and Mom answered his knock. "Why, John, come in!" she exclaimed, evidently pleased to see him. I was sprawled on the rug in the living room, reading *The Voyage of the Space Beagle* by A. E. van Vogt. He had written a lot about the rocket his heroes were voyaging in, but nothing on how it worked. That was a disappointment.

"Elsie," Mr. Dubonnet greeted her grimly, taking off his black helmet. He stayed on the porch. "Is Homer home?"

Dad was in the kitchen, probably getting an apple. After his cancer was cut out of him, the doctor had prescribed as many apples as he could stand, and he ate a lot of them. Dad came to the front door. "You want to talk to me, Dubonnet, come see me at my office." He said it in as mean a tone of voice as I ever heard him use.

"What's gotten into you, Homer?" Mom gasped. "Please come in, John!"

The union man stood his ground. "It's okay, Elsie. Homer, could you step outside? We need to talk before I go down to the union hall for the meeting."

Dad frowned, but went out and closed the storm door behind him. I couldn't hear what was being said between him and Mr. Dubonnet, but I got up and came into the foyer so I could watch. Mom gave me a disapproving look and I retreated back to the living room, carefully positioning myself to where I could still see what was going on. I remembered that Mom, Dad, and Mr.

Dubonnet had all come out of Gary. I also recalled that Mr. Dubonnet had been the valedictorian of their class and a star football player. It had never exactly been said, but I think Mr. Dubonnet had even taken my mom out a few times way back then. After a while, Dad opened the storm door to come back inside. "The company's given you a good job, a house, and a decent life, Dubonnet," he was saying, "and all you want to do is tear it down."

"The cutoff was not done properly according to our contract," Mr. Dubonnet said reasonably. "You know that, Homer."

Dad put his hand on the doorknob. "The company did what it had to do."

"How they turned you into such a company man I'll never figure," Mr. Dubonnet said. This time his voice was hard and bitter.

"Better than to throw in with a bunch of John L. Lewis commies!" Dad shot back.

Mr. Dubonnet shook his head. "The trouble with you, Homer, is that you don't know who your real friends are. When the company gets into trouble, it'll throw you out like a dead mouse."

Dad stepped back onto the porch. "And the trouble with you, Dubonnet, is you can't get over I got the Captain's job." Dad was about to say more when he started to cough and grabbed his chest.

"That's it, Homer," Mr. Dubonnet chided him, "cough your lungs out. You might be the mine superintendent, but you've got the common miner's disease."

"Stop it, both of you!" Mom spat.

"Stay out of this, Elsie," Dad gasped and took a big, strangled breath.

"Look at him," Mr. Dubonnet said to my mother. "You think the company cares anything about his lungs or anybody else's? Hell, no! This is what the great Captain did for us with his continuous miners."

Dad shook his head and searched for air. "You lay off the Captain," he gasped. "He was a great man. I've got allergies, that's all. Look at my daddy. Look at yours too. They worked in the mine all their life, and they never had any problems with their lungs."

"Our daddies dug coal out with picks, Homer," Mr. Dubonnet said, back to being calm again. "The continuous miners grind the coal up, fill the air with dust. After we get this cutoff settled, it's the next thing I want to talk to you about. We need some way to protect the men from the dust."

"I'll thank you to get off my porch," Dad choked.

"John, maybe it would be best," Mom said softly, putting her hand on Dad's arm. He shrugged it off.

Mr. Dubonnet put his helmet back on. "Elsie, you're a fine woman. I always thought you deserved better." He turned and went out the gate and walked across the street toward the gas station.

Dad lurched back inside and slumped into his easy chair. "Damned union John L. Lewis sonuvabitch," he muttered. "Still thinks he's the great football player. Well, I could've played, but I had to work, picking coal at the tipple after school."

"I know that, Homer," Mom said, watching him from the foyer. Her gentle tone surprised me.

Dad's hands were trembling as he reached for his paper. "You're a good woman, Elsie," he said.

"I know that too, Homer," she replied softly.

"You could've had your pick."

"I did." She looked at me, probably just noticing I was in the living room. "Go to your room," she barked. "Study!"

I nodded and went up the stairs, two steps at a time. Outside, a line of cars rumbled into the gas station, and I looked out the window to see what was going on. Mr. Dubonnet got into one of the cars and then they all headed down Main Street. I guessed they were headed to the union hall.

J UST before Thanksgiving, Doc Lassiter ordered Dad to get an X ray. When he refused, Doc went to Mr. Van Dyke, the only man in town who could tell Dad what to do. Mr. Van Dyke was the mine's general superintendent, a courtly man with silver hair who had been sent by the steel company to keep watch on its holdings. Dad, a company man through and through, had no choice but to comply with Mr. Van Dyke's orders. He went over to Stevens Clinic in Welch. When he came home, I was upstairs in my room, reading. "A spot," I heard him tell Mom. "About the size of a dime."

"Oh, Homer, for God's sake," I heard her say in a small and worried voice I hadn't ever heard her use before. "What are you going to do?"

"I'm not going to do anything," he replied blandly. "Why're you looking at me like that? Don't worry about it. I'm only telling you because you'll find out anyway. The only way to keep a secret in this town would be if I tore down every backyard fence in it."

Dad went into the living room and sat down and snapped open the *Welch Daily News*. Mom looked up and saw me. She scowled, crossed her arms, and went back to her kitchen. She was soon rattling pots and pans. I went back to my room and stared at nothing, feeling a little panicky. I'd been around Coalwood long enough to know miners with spots on their lungs were supposed to quit the mine. It was not unusual to see them sitting on the Big Store or post office steps during the day, quietly hacking up black spit. For some reason, if it was their lungs that made them quit, they were allowed to stay in Coalwood as long as they could afford the rent for their houses. But I never imagined the common disease of the mine could affect my dad. He seemed far too tough for that. He had a spot on his lungs the size of a dime. I'd have to ask somebody about how bad that was. Maybe Roy Lee. His older brother worked up around the face where the coal dust was the thickest. Yes, I would ask Roy Lee. He would know.

IN December 1957, the United States made its first attempt to put a satellite into orbit with its *Vanguard*. I saw the result on television. *Vanguard* managed three tentative feet off the pad, lost thrust, and then blew up. According to the papers, the whole country was shocked and disappointed. I was too. I read some newspaper editorials and listened to television commentary that wondered if perhaps western civilization itself might soon be at an end with the technologically superior Russians taking over. I might have worried even more about *Vanguard*'s failure if I hadn't had rocket problems of my own. *Vanguard,* after all, had managed to fly a yard higher than I had. There were a lot of smart people working on that project, and I guessed they'd figure things out, sooner or later. I, on the other hand, was all alone. That's why I decided, like it or not, I had to talk to Quentin.

5

QUENTIN

QUENTIN WAS THE class joke. He used a lot of big words often delivered in a pseudo-English accent, and he carried around an old, cracked-leather briefcase, stuffed to overflowing with books and who-knew-what-else. While the rest of us played dodgeball or did calisthenics in phys. ed., he always had some excuse not to participate—a sprained ankle or a headache or some such—and sat in the bleachers and read one of his books. While all the other students traded gossip and nonsense in the auditorium in the morning and at lunch, Quentin always sat alone. He had no friends as far as I could tell. Although everybody, including me, made fun of him, I was pretty certain he was some sort of a genius. He could expound on nearly any subject in class until the teacher had to ask him to stop, and if he'd ever made less than a hundred on a test, I wasn't aware of it.

I figured if there was anybody who might know how to build a rocket, it was Quentin. The next morning before classes, I sat down beside him in the auditorium. Startled, he pulled his book down from his face. "I don't let anybody copy my homework," he said suspiciously.

"I don't want to copy your homework," I replied, although I would have taken his algebra if he had offered. "Do you know anything about rockets?"

A little smile crossed his face. Quentin wasn't a bad-looking kid for a genius. He had a narrow face, a sharp nose, crisp blue eyes, and jet-black hair that looked as if it had been plastered down with about a quart of Wildroot Cream Oil. "I wondered how long it would take for you to come to me with that question. I heard about your rocket, old boy. Blew up, did it? What made you think *you* could build a rocket? You can't even do algebra."

"I'm getting better," I muttered. It was amazing to me that everybody, even Quentin, knew my business.

"One of my little sisters can already do algebra," he advised me. "I taught her. It's really quite simple."

In less than a minute, he'd already pretty much irritated me. "So what do you know about rockets?" I asked him. "Anything?"

"I know everything," he replied.

He had said it too easily. "Let's hear it, then," I said, doubtfully.

He raised one of his bony shoulders in a shrug. "What's in it for me?"

"What do you want?"

"To help you build the next one."

That was a surprise. "If you know so much, why don't you build your own?" I demanded.

Quentin built a little church with his hands. "I've considered building a rocket for some time, if you must know. Practical reasons, unfortunately, have prevented me from taking action. It takes teamwork to build a rocket, and materials. My observation of you is that you have certain . . . leadership abilities that I do not." He locked his eyes on me. They were intense, almost like they were capable of shooting rays. "The other boys will follow you," he said. "And you, being the Coalwood superintendent's son, can probably get all the materials you need."

His ray-gun eyes made me want to look away, but I didn't. "What's your angle?" I demanded.

"Ho-ho!" he exclaimed. "The same as you, old chap! If I learn how to build a rocket, I'll stand a better chance of getting on down at the Cape."

"First you've got to go to college," I reminded him.

"I'll go to college," he said resolutely. "But it won't hurt to get some good practical rocket-building experience under my belt." He put out his hand. "How about it? You want to team up?"

It was the best offer I'd gotten since I'd started my rocket-building career, but I was still a little reluctant. I didn't have much of a reputation at Big Creek, but it was still better than Quentin's. When I didn't respond to his hand, he grabbed mine and shook it, big strokes up and down. I hastily pulled my hand away and looked around to see if anybody had noticed. I knew the football boys would accuse me of holding hands with Quentin if any one of them had seen it.

"So what do you know?" I demanded, my face flushed with the potential embarrassment of it all.

"Calm down, old chap," he said. "All shall become perfectly clear." He leaned back and took a deep breath and then began to talk just as if he was reading something straight from a book. "The Chinese are reputed to have invented rocketry. Something called 'Chinese arrows' are mentioned in Europe and the Middle East as far back as the thirteenth century. The British used rockets later aboard their warships during the Napoleonic Wars and the War of 1812. That's where 'the rocket's red glare' comes from in 'The Star Spangled Banner.' Then there was the Russian Tsiolkovsky, Goddard the American, and von Braun, of course. Each of them added to the body of rocket knowledge. Tsiolkovsky was a theorist, Goddard applied engineering principles, and—"

I stopped him. "I don't need to know this stuff. I need to know how a rocket works."

Quentin cocked his head. "But that's so *elementary*. Newton's third law. For every action, there is an equal and opposite reaction."

I remembered from one science class or another something about Newton, but I couldn't put my finger on his laws. "How do you know that?"

"Read it somewhere."

"Read it *where*?"

Quentin frowned, disturbed at my attempt to cut through his bull. "A physics book, I suppose," he said stiffly. "Can't exactly say which one. I thumb over to Welch every Saturday to the county library. I tend to pick out random shelves and just read every book on it until I'm done."

I could see it was going to be necessary to be more specific with Quentin. "What kind of fuel does a rocket use?"

"The Chinese used black powder."

"Black powder?"

He looked at me carefully, as if to determine if I was joking. "Black powder. It contains potassium nitrate—saltpeter, you know—and charcoal and sulfur."

Saltpeter? Quentin sighed and then explained in some detail the chemical's properties. It was an oxidizer, which, when combined with other chemicals, produced heat and gas, necessary to make a rocket fly. "It can also kill you down there," he finished up, pointing at his crotch.

"What do you mean?"

"It fixes men so they can't . . . you know."

"What?"

Quentin flushed. "You know." He straightened a crooked finger. "That."

"Really?"

"Well, that's what I read."

I thought I'd better get back to rockets. "Where can I buy some of this black powder?"

"You can't buy it, far as I know," he said. "You've got to mix it up. Saltpeter, sulfur, and charcoal, that's what we need. Can you get it?"

I wasn't certain, but I wasn't going to let him know it. "I'll get right on it."

Quentin grinned broadly and suddenly started to rattle on like I was his best friend. He opened up his briefcase and showed me all the books within, most of them on general science, but one of them a novel titled *Tropic of Cancer.* "You want to know about girls? This is the one," he said slyly.

"I already know about girls."

He tapped the book. "No, you don't."

When the bell rang and we stood up, I noticed for the first time Quentin's worn, faded shirt, thin at the elbows, and his patched cotton pants and his scuffed, ankle-high shoes. Quentin wasn't a Coalwood boy. He came from Bartley. He was one of the kids my mother told me to notice. The mine at Bartley was always having cutoffs and strikes, and in the last few years, many Bartley families had slid into poverty and misery. Quentin's father was probably out of a job. In 1957 southern West Virginia, you wouldn't likely starve if you didn't have any money. There was always bread and commodity cheese you could get from the government. But that was about all there was.

Roy Lee stopped me in the hall. "What were you talking to that moron Quentin about? And did I see you holding his hand?"

I was peeved enough at Roy Lee to not answer, but then I figured it would aggravate him more to tell the truth. "We're going to build a rocket, him and me."

A surge of kids passed us, Dorothy Plunk among them. "Hi, Sonny, Roy Lee," she called angelically. I opened my mouth, but nothing came out.

Roy Lee shook his head and leaned against the lockers. "Gawd almighty. You want to stop all chance of ever having any kind of social life? Dorothy Plunk sees you and Quentin hanging out together, she's going to lose interest right there."

I looked after her, trying not to stare at her cute little bottom swinging back and forth down the hall. "Dorothy doesn't care anything about me, anyway," I said, sort of breathless.

Roy Lee didn't try to hide what he was looking at. He watched Dorothy all the way down the hall. "Whew," he whistled. He pulled his eyes back to me. "You got good taste anyway, Sonny. Why don't you ask her out? Double date with me next weekend. We'll go parking at the Caretta fan."

"She'd just say no."

Roy Lee shook his head again as if I was the burden of his life. "If you don't ask her out, I will."

"You wouldn't do that!" I bleated.

He wiggled his eyebrows at me, his crooked grin clearly salacious. "Oh, but I would!"

Roy Lee had me in a box. If he asked Dorothy out and she went, I wasn't sure I could live through it, knowing what Roy Lee was likely to try with her. And what if she did it—or if Roy Lee claimed she did? My life would be pretty much ruined forever. I had no choice. I chased after Dorothy, catching her with Emily Sue at the entrance to biology class. "I'm sorry I got sick over the worm," Dorothy said, first thing.

"Dorothy," I said, my heart pounding in my ears, "would you like to go to the dance with me Saturday night? With Roy Lee? I mean, in his car? I mean—"

Her big blues blinked. "But I already have plans!"

The blood drained from my face. "Oh . . ."

"But if you'd come over to my house on Sunday afternoon," she purred, "I'd love to study biology with you."

To her house! "I'll be there!" I swore. "What should I bring? I mean—"

"Just yourself, silly." She looked me over, studying me, and I kind of thought she liked what she was seeing. "We'll have so much fun," she concluded.

Emily Sue had been observing all this. "Be a little careful with this one, Dorothy," she said.

"Whatever do you mean?" Dorothy asked her friend.

They talked to each other as if I weren't even there. "Sonny's nice," Emily Sue said succinctly.

"Well, so am I!" Dorothy said back. She went on into the classroom.

Roy Lee had been loitering nearby, listening. He came up and stood beside Emily Sue. "What do you think?" he said.

They were talking as if I weren't there too. "Dangerous," Emily Sue told him. "But probably not fatal."

During class, I couldn't help but sneak looks at Dorothy at her desk while she worked on a drawing of frog intestines. She had the adorable habit of letting the pink tip of her tongue protrude from her full, delectable lips while she concentrated. She was wearing a white pinafore blouse with a blue ribbon around its collar, which made her look so very innocent, yet the way she filled the blouse out troubled me with indecent thoughts. She caught me looking once and gave me a demure little smile while I blushed. I couldn't figure out how so much perfection could wind up in one person. Then a little misery inserted itself. If Dorothy had plans for Saturday night, I didn't imagine it was to bake cookies with her mother.

THERE was a company-store system in most of the towns in southern West Virginia. They usually featured easy credit and inflated prices. If a miner got into enough debt with a company store, the company stopped paying the miner with U.S. dollars and issued his pay in the form of *scrip*—company money good only in the company store. It was an insidious system. A popular song across the country in the late 1950's was Tennessee Ernie Ford's "Sixteen Tons," where he sang about a miner owing his soul to the company store. That was just about the truth for a lot of West Virginia miners.

As part of his social agenda, the Captain abolished the worst aspects of the company-store system in Coalwood. He brought in a college-educated manager—Mr. Devotee Dantzler, a Mississippi gentleman—to make certain prices were kept fair and no miner was gouged. The Captain dictated that credit could be given, when necessary, but the books were to be watched closely. No miner was allowed to get himself too far into debt. Scrip in Coalwood was issued sparingly. Smaller stores were built around the town for the convenience of the population. Under Mr. Dantzler, the Big Store became a source of town cohesiveness and a social gathering spot.

The Big Store contained a little bit of everything: hard-toe boots, leather utility belts, helmets, coveralls, and the cylindrical lunch buckets the miners favored; clothes for the whole family, groceries, and umbrellas; refrigerators, baby carriages, radios, and television sets with free installation onto the company cable system; pianos, guitars, record players, and a record department too. It had a drugstore where you could get Doc's prescriptions and a wide variety of patent medicines, and a soda fountain where you could get pop and candy and a milk shake so thick a spoon would stand straight up in it. It had auto parts and lumber; shovels, picks, rakes, and seeds for the little gardens the miners scraped into the sides of the mountains. It even had a limited choice of coffins, hidden away in a back room. It was technically illegal to bury anyone on company property, but the colored people had a cemetery somewhere up Snakeroot Hollow. My father, and the company, looked the other way on it.

The Big Store had just about everything anybody in Coalwood needed, but would it have rocket fuel? With my cigar box of dollars and scrip left from my defunct newspaper delivery business, I went to Junior, the clerk at the drugstore counter, to find out. Junior was a rotund little man with a cherubic face, who was as smart as a whip and liked all over the town. When Junior worked on the store truck that delivered heavy things (like refrigerators) in the afternoon, he was welcome to come right into anybody's house, even though he was a Negro. Most of the ladies loved him, even petted on him a little. He rarely got away from them after a delivery without some tea or coffee and cake. I saw him once in Mom's kitchen, admiring her mural. Mom was beaming. It was rumored that Junior had once attended college, which put him ahead of even my father. Junior heard my order and cocked his head doubtfully. "Saltpeter?" he demanded in his raspy voice. "Your folks sent you after that?"

"It's for me," I said forthrightly. "Science project. I need sulfur and charcoal too."

Junior adjusted his wire-rimmed spectacles and seemed to make a mental calculation. Then he went into the back and brought out a can each of sulfur and saltpeter and a ten-pound bag of cooking charcoal. "Listen, rocket boy," he said. "This stuff can blow you to kingdom come. I think you know what I'm saying."

I mumbled, "Yessir," and paid with scrip. I loaded my treasures on my bike and rode home. As I passed a line of miners walking to work, Mr. Dubonnet hailed me down. "I hear you're going to build another rocket," he said.

"Yessir. I'm thinking about going down to Cape Canaveral and joining up with Wernher von Braun."

He seemed to brighten at the news. "That's a good thing. You're too smart to stay here."

The line of coal cars beside us suddenly began slamming against each other, just before being pushed to the tipple. It was as loud as a hundred car wrecks happening all at once, but no one, including me, even bothered to look in its direction. We heard that sound every day. "Mr. Dubonnet, you're smart too," I said in an attempt to figure out why everybody lately seemed to want me out of town. "After the war, why did you come back to West Virginia if it's so bad?"

He laughed. He had a deep, rich, ho-ho-ho kind of laugh that was wonderful to hear. "You got me there, Sonny." He began to walk and I pushed my bicycle beside him. "I guess these old mountains, the mines, the people get in your blood," he said. "When I got back from overseas I couldn't wait to get home to McDowell County. It's where I belong."

There was the thing. He'd hit right on what I'd been wondering about since Mom's backyard lecture. "How do you know I don't belong here too?" I wondered.

He stopped and raised his eyebrows as if I had said the most amazing thing. I guess my ignorance was a continuing surprise to everyone in Coalwood. "Well, you do, of course," he answered. "Anybody raised here belongs here. You can't belong anywhere else."

The empty coal cars shrieked as the locomotive, perhaps a mile down the track, began to push them toward the waiting tipple. I shouted to be heard. "Then I don't understand why I'm supposed to leave!"

He stopped again, the other miners passing us by at a hard slog, the shift-change hour approaching. "Don't you understand?" he yelled. "In just a few years, all this will be gone, almost like it never existed." The coal cars began to roll and the noise lowered to a deep rumble. Mr. Dubonnet lowered his voice with them. "Even the union can't put the coal back in the ground."

I knew I probably shouldn't ask him anything about my father, considering the row I'd observed between them, but I couldn't resist. "Does my dad know this?"

Mr. Dubonnet grimaced. "He knows. But he acts like he don't."

"How come?"

"Now, that's something you should ask him," Mr. Dubonnet said, his face turning as hard as concrete. "Good luck with your rockets, Sonny." He joined the line of men and quickly disappeared, one black helmet in a river of black, bobbing helmets, all going up the path to the tipple. I looked back down the valley at all the houses. Women were out on their front porches with their mops and buckets, waging their never-ending battle against the coal dust. The coal cars kept trundling past until a big black steam locomotive, puffing huge gouts of white smoke, finally appeared. It churned past, its engineer giving me a wave. I waved back distractedly. With all this activity, I just couldn't imagine it ever ending. Maybe Dad and I had a similar blind spot.

Beside the washing machine in the basement was a wide counter and a deep steel sink. I had decided it would be my rocket laboratory. As soon as I set my chemicals on the counter, the upstairs door opened. "Sonny?" Mom called, and I yes-ma'amed her. "Remember what I said. Don't blow yourself up."

News in Coalwood traveled a lot faster than a boy on a bike.

QUENTIN hitchhiked over on Saturday and I introduced him to my mother. He made a short bow at the waist to her, an Errol Flynn kind of move I'm sure he had seen in the movies. Mom was impressed by it, however, her hand going to her mouth almost like a bashful girl's. She rarely baked cookies, but I soon smelled their aroma drifting down from the kitchen. When she thumped down the basement steps with them and two glasses of milk, the plate she handed Quentin was piled twice as high as mine. "These are marvelous, undoubtedly the most delicious cookies I have ever tasted in the entire history of my life," Quentin told her after a nibble. Mom looked tickled. She wondered what else she could do for us.

"*Nothing*, Mom," I answered. I just wanted her to leave so we could get to work.

She seemed to want to loiter. "If you need anything, just call me."

"We will, Mom. See you later, Mom."

After my mother had gone back to the kitchen, Quentin worked on his cookies for a while while I waited impatiently. Finally, he took a final swig of milk, wiped his mouth on his sleeve, and picked up the bag of saltpeter. He looked inside it. "Looks pure," he said. I wondered how he would know.

With Dandy and Poteet watching us furtively from a dark nook beside the coal furnace, we began to work. First, we mixed up several small batches of what we hoped was black powder and, as a test, opened the grate and threw a spoonful of each into the coal-fired hot-water heater beside the washing machine. The ingredients hissed feebly, but it was impressive enough for the dogs to beg to be let out. I opened the basement door and they bolted outside. "What do you think?" I asked. Quentin shrugged. Neither of us knew how rocket fuel was supposed to burn.

We decided to test two of our best mixtures inside devices we hoped resembled rockets. There was some one-inch-wide aluminum tubing under the back porch that Dad had brought home from the mine to make a stand for Mom's bird feeders. I appropriated it with a clear conscience since it looked as if he were never going to get around to it. I hacksawed off two one-foot lengths. Quentin called the lengths our "casements." We hammered in a short length of broom handle at one open end and then poured in our powder mixes, crimping the other end with pliers to form a constriction the *Life* magazine diagram called the rocket "nozzle." The result was obviously crude, but it was for testing purposes only. We attached triangular cardboard fins with model-airplane glue. We knew the fins would probably burn off, but they would at least give our rockets something to sit on. "We need to see how the powder acts under pressure," Quentin said. "Whatever the result, we'll have a basis for modification."

I was becoming used to Quentin's way of putting things. What he was saying was that we had to start somewhere, either succeed or fail, and then build what we knew as we went along. It seemed to me, considering all the rockets that I read about blowing up down at Cape Canaveral, that was the way Wernher von Braun and the other rocket scientists did their work too. Without Quentin, I might have been too embarrassed to fail in front of God and everybody. With him, no matter what happened, I felt "scientific." Failure, after all, just added to our body of knowledge. That was Quentin's phrase too. *Body of knowledge.* I liked the idea that we were building one.

After the fins dried, I decided we would test our creations behind my house, over by the creek. I didn't think there was anything over there anybody would care about if we blew it up. To my surprise, Roy Lee appeared, claiming he just happened to be in the neighborhood. I think he'd actually been hanging around waiting for me and Quentin to come out.

The first rocket emitted a boil of nasty, stinking, yellowish smoke and then fell over, the glue on its fins melted. "Wonderful," Roy Lee muttered, holding his nose. Quentin silently wrote the result down on a scrap of notebook paper. *Body of knowledge.*

The second rocket blew up. A good-size chunk of shrapnel twanged off the abandoned car we were hiding behind. A cloud of oily smoke covered us. Dad came out on the back porch and yelled, "Sonny! *Get over here right now!*" Obediently, we followed the smoke, reaching him as it did. He wrinkled his nose. "Didn't I tell you not to do this again?"

I didn't get a chance to answer. Mom came out. "Homer, telephone." She gave us boys a little smile while she waved the smoke away from her.

Dad went after the call and then came back out on the porch. He ignored Quentin and Roy Lee, his eyes on me. "As soon as I put the phone down, it rings again. People are complaining about the stink and smoke. I want this stopped. Do you understand me?"

Mom quickly amended his meaning. "Not behind the house, dear. You need to find a better place."

Dad turned on her. "Elsie, they've got to stop trying to burn Coalwood down!"

She kept her smile on us boys. "Okay. I'll make them promise. You won't burn this wonderful, beautiful city down, will you, boys?"

"No, ma'am!" we chorused.

"You see?"

Dad stared at her and then shook his head and went inside. She followed him, leaving us boys to contemplate what had been, after all, our scorched, stinking failures. Quentin finished his notes. "First sample was too weak, the second too strong," he said. "Now we know where we are. This is good, very good."

Across the creek, some younger children gathered—dirty, snot-nosed urchins all. "Hey, rocket boys! Why don't your rockets fly?" they chorused.

Roy Lee picked up a rock and they scattered, giggling.

I HITCHHIKED to War on Sunday afternoon. Dorothy's house was across the railroad tracks on the mountain that overlooked the town. Her mother welcomed me with a delighted grin, as if she never wanted to see anybody more in her life. I could see a little of Dorothy in her face, but, unlike her daughter, she was a big, robust woman. Although Dorothy's hair was a sandy color, her mother's hair was the color of an orange. Dorothy's father, a lanky, nearly bald man, stepped in from the kitchen and listlessly shook my hand. The owner of a gas station in War, I could tell he was used to Mrs. Plunk doing most of the talking. Both parents disappeared into the kitchen, leaving me and Dorothy in the living room with our biology books. As it turned out, we didn't study much. She wanted to know all about my rockets. "I'm so proud just to know somebody who does something so interesting!"

Emboldened, I told her I was going to try to learn as much as I could and go down to Cape Canaveral and join up with Wernher von Braun. "Oh, Sonny," she said, "I know you're going to be an important person someday. When you get to Florida, will you write me and tell me all about it?"

I struggled to find the nerve to tell her I didn't want to write her, that I wanted her to be there by my side. But before I could find my voice, she said, "I want to be a teacher and a mom, the best one there ever was. I so love children—"

"So do I!" I exclaimed, although it was news to me. If Dorothy wanted it, I did too.

We continued talking, about friends and our parents. I told her about my mother—all the little funny things she did, about Chipper, her squirrel she kept in the house, and her mural on our kitchen wall. When I described Dad, all I could say was that he was in charge of the mine and worked a lot at it and, yes, he had caused the suit to be filed on behalf of the Big Creek team. "What's it like to be Jim's brother?" Dorothy asked, even though I hadn't mentioned him.

I really had never given that particular subject much thought. "Okay, I guess" was the best I could do.

"He's such a *good* football player!"

I shrugged. "Uh huh . . ."

"I think you're much more interesting," she said.

That brightened me up. It seemed like the right moment to ask her out on a date. "Dorothy, you know Roy Lee has a car, and I was just thinking that maybe you and me—"

"Do you know what, Sonny?" she interrupted. "I've never been outside of West Virginia. Isn't that sad? How about you?"

Her question caused my own to die on my lips. I told her I had been several times to Myrtle Beach, South Carolina. Mom loved it there. And Dad had driven the family to Canada when I was in the third grade, all the way to Quebec.

She seemed thrilled. "Tell me about Quebec."

I remembered how clean everything was. The French language had also made an impression. "It sounded real pretty to hear it," I told her.

"Someday I'll go there and hear it too," Dorothy said solemnly.

I was halfway home before I realized Dorothy had diverted me from asking her out. I resolved to do it the next morning. I scanned the auditorium and found her with a group of her girlfriends huddling around a trio of senior football players. Dorothy was wearing a tight pink sweater and a black poodle skirt and was on her knees in the chair in front of them, her hands covering her mouth as she laughed at something one of the boys had said. I edged in beside her and stood awkwardly while she bantered back and forth with him. "Saturday night then?" he asked, and she nodded eagerly.

"Oh, hi, Sonny!" she said brightly and then slipped past me, to join her future date on a stroll up the aisle. I just stood there, my heart sinking to my toes.

6

MR. BYKOVSKI

Auks I–IV

ON JANUARY 31, 1958, the Army Ballistic Missile Agency (ABMA), led by Dr. von Braun, was ready to launch the *Explorer-1* satellite aboard a *Jupiter-C* rocket. It was to be a night launch, so I stayed up to watch television, hoping for good news. Around 11:00 P.M., a bulletin interrupted the *Tonight Show* with an announcement that the launch had been a success. Film of the launch was promised momentarily. I started a vigil, lying on the rug in front of the television set, staring at the set, which displayed nothing but a sign stating STAND BY. Mom, Dad, and Jim had long since gone to bed. Daisy Mae joined me on the rug, curling up behind me in the bend of my knees. The bitter cold outside had also chased our old tom Lucifer in, and he was curled up in Dad's easy chair. It was good to have them as company. I reached back and patted Daisy Mae's head. "Good old girl," I told her. "Good old cat." She rewarded me with a purr and a lick on my hand.

Daisy Mae was a pretty cat, a fluffy calico, and was special to me. Four years earlier, when she had wandered in from the mountains, I hid her for a day, secretly feeding her in the basement. When Mom discovered her, she said I'd have to find the kitten another home, pointing out we already had two dogs, a squirrel, and a cat, and that was enough animals. After I pouted about it for a day, Mom gave in. "If you want this kitten," she said, "you'll have to take care of her." I readily agreed, easy enough to do (the agreeing part). Daisy Mae had kittens right off, a pretty litter quickly snapped up by the neighbors. By then, Mom had completely adopted her into the family and, as I knew she would, took care of her as she did all the other animals, feeding her and spending hours picking fleas out of her coat. Mom thought Daisy Mae was such a pretty but delicate cat that she decided we'd have her fixed. To my knowledge, no other cat or dog in Coalwood had ever been neutered before. Mom drove Dad's Buick with me holding Daisy Mae on my lap all the way to the veterinarian over in Bluefield, forty miles and six mountains away. It was the first time any of our animals had ever seen a vet. After she healed, Daisy Mae became even more loving, waiting for me to come home from school and sleeping on my bed at night. I often talked to her before I went to sleep, especially when I was frightened, or worried. She was just a comfort when everybody else in the family seemed at odds. Of course, I never told anyone else I talked to my cat, certainly not any of the other boys. I'd have never lived it down.

Around midnight (it was a Friday and not a school night), I was surprised by a knock on the front door, and in came Roy Lee, Sherman, and O'Dell to join me. They bedded down on the couch and the floor. We talked some, mostly about girls, but then O'Dell and Sherman kind of drifted off. I'd been meaning to ask Roy Lee about the spot on Dad's lung, so I took the opportunity. He tucked himself in the corner of the couch and gave me a worried look. "I'll ask Billy," he said. Billy was his brother.

"Don't tell him why. Dad doesn't want anybody to know."

Roy Lee gave me a funny look. "Sonny, I already knew. I guess everybody in Coalwood knows."

I put my head down on the rug and pretty soon I went to sleep. I woke during the night, finding the picture on the television turned to snow. I kept waking up and falling back to sleep. At dawn, I was awake when the picture flickered back on and an announcer said to stand by. I woke the others up and then, without preamble, film of the launch was run. Dr. von Braun's rocket lifted off the pad in a caldron of fire and smoke and went right up into the night sky without a moment of hesitation. We whooped and cheered at the sight of it. O'Dell got up and did a little jig and then fell back on the couch and put his feet up in the air and made like he was riding a bicycle. I wasn't so demonstrative, but I felt proud and patriotic. Dad came downstairs, let Lucifer and Daisy Mae out, and found us boys clustered around the set. He looked us over. "Did it work?"

It was the first time I remember him ever expressing any interest in space. "Yessir!" we roared.

He stared at the television, where Dr. von Braun's rocket kept taking off again and again. "I don't know what to make of it," he said. I'd never heard him say anything like that before.

"We're going into space, Dad," I said, by way of an explanation.

"Little man," he replied, "in your case, I think sometimes you're already there." I took that as a compliment and beamed. He looked back at me with his eyebrows raised.

Mom appeared in her housecoat. She smiled drowsily at me and the other boys. "Did it work?"

"Yes, ma'am!"

"I think that's wonderful. Don't you, Homer?"

Dad had gone to the kitchen. "Wonderful," he said, his voice afar.

Mom looked us over. "You boys want some breakfast? How about some waffles?"

"Yes, ma'am!"

Later that same day, I gathered Roy Lee, Sherman, and O'Dell in my room. "Okay, here's what we're going to do," I said.

Roy Lee fell back on the bed and groaned. "Every time you say that, we always end up in trouble."

I laid out my plan. I was forming a rocket club to be called the Big Creek Missile Agency (BCMA), named in imitation of von Braun's ABMA. Quentin and I were going to be in it. We were going to learn all there was about rockets and start building them. This was to be a serious thing, not playing. If the others wanted to join us, they were welcome. I figured Roy Lee would get up and walk out rather than belong to anything with Quentin in it, but instead he sat up on the bed and rubbed his chin thoughtfully. "Sonny, I like it. It sounds like fun. Count me in." I think he was inspired by the success of the *Explorer*. Sherman and O'Dell readily agreed too.

"The Big Creek Missile Agency is hereby formed," I said. "I'm the president. O'Dell, I'd like for you to be the treasurer and in charge of supplies. Roy Lee, because you've got a car, we'll need for you to handle transportation. Sherman, if you'd take care of publicity and setting up our rocket range, I'd appreciate it. Quentin is going to be our scientist. Any questions?"

Roy Lee said, "Any girls in this club, or do you have to have a rocket in your pocket?"

"Or in your case a pencil," O'Dell jeered at Roy Lee.

"You oughta know," Roy Lee replied, his eyebrows dancing. O'Dell blushed. Trading insults with Roy Lee was never a good idea, even for a bright kid like O'Dell.

"Where's our rocket range going to be?" Sherman asked me.

"We'll have to think on that," I said.

"There's an old slack dump up behind the mine," Sherman said. "That might do."

Slack was the tailings of the mine, coal with too much rock in it. Wherever it was dumped, nothing grew. I thought Sherman had a good idea. "We'll try it," I agreed.

"So what do we do now?" O'Dell asked.

"We build a rocket."

"How?"

"Got to work on that," I admitted.

After we finished our meeting, not deciding anything else except what time we were going to meet the following weekend, the boys went home. I stopped Roy Lee at the door. "Don't ask your brother about Dad's spot," I said.

Roy Lee nodded. "You don't want to know how bad it is?"

"No, I don't." That about summed it up. I couldn't do anything about it anyway.

AT lunch during the following days, Quentin and I worked on how to build a rocket, sketching out crude drawings and theorizing. We were proceeding mostly by instinct. Despite a search from top to bottom at the McDowell County Library, Quentin still couldn't find any books to help us. While we worked, both of us ate out of my lunch bag. He told me he usually skipped lunch because eating too much was unhealthy. I noticed, however, that his health regimen didn't keep him from eating more than half of my food. When I mentioned this to Mom, she started putting in an extra sandwich because, she said, "You're a growing boy." I wasn't fooled. She might as well have written QUENTIN on it in big capital letters.

One day, on our way to class after lunch, Quentin and I were walking past the Big Creek football trophy case, just outside the principal's office, when he stopped and put his hand on the glass. "Maybe one day we'll have a trophy in here, Sonny, for our rockets."

"Are you kidding?"

"Absolutely not. Every spring, science students present their projects for judging at the county science fair. If you win there, you go to the state and then to the nationals. Big Creek's never won anything, but I bet we could with our rockets."

Quentin and I saw their reflections in the case when they came up behind us—Buck and some of the other football boys, looking huge in their green and white letter jackets. "What the hell you two morons doing in front of our trophies?" Buck demanded. He squinted past us. "Oh, no! Is that your filthy handprint on our trophy case?"

"Let's murder these sisters," a tackle snarled. A growl of agreement rose from the assembled giants.

We turned to face them. "I assure you chaps—" Quentin started to explain.

"*I assure you chaps!*" Buck mocked Quentin. "You really are a little sister, ain't you?" He bulled his face in close to us, his chin prickly with whiskers. There was a brown chewing-tobacco stain in the lower left corner of his mouth. I could smell its sweetness on his breath. "I assure you I'm gonna kick your chapped tails. You especially, Sonny. I still owe you, big time."

Jim came by, his latest girl on his arm. He eased her on down the hall and came over to see what was going on. He saw it was me and said, "Leave them alone, Buck."

Jim could take him apart and Buck knew it. "I wasn't going to hurt your little four-eyed sister moron brother," Buck said, lying through his teeth. "But this little sister," he said, nodding at Quentin, "I'm going to kick his tail."

"You can kick both their tails for all I care, but do it somewhere else," Jim said, dispelling any thought I might have had that he cared anything about me. He nodded toward the principal's office. "I just don't want the team to get into any trouble."

Mr. Turner strutted out of his office at that moment. A young woman was with him. I recognized her as Miss Riley, a Concord College senior assigned to Big Creek as a student science teacher. If what I heard was correct, she would be teaching us chemistry next year. Mr. Turner was a banty-rooster kind of man who kept the entire school under his thumb. He took one look at the assembly in front of the trophy case and said, "If this hall isn't cleared of boys with letter jackets in two seconds, I know who won't be playing football anymore."

Jim and Buck and the football players disappeared, almost as if they got sucked up into the ceiling, leaving Quentin and me standing exposed. Mr. Turner looked us over. "Are you two boys plotting something nefarious?"

Quentin was frightened into honesty. Besides that, he understood what *nefarious* meant. "I was just telling Sonny," he said, "I think someday there will be a trophy in here for the Big Creek Missle Agency."

Mr. Turner frowned deeply. "And what, pray tell, is the Big Creek Missile Agency?"

"Our rocket club," I said when Quentin hesitated.

He looked at me closely. "Mr. Hickam, isn't it? Jim's brother? Did I not hear that you blew up your mother's rose-garden fence? That sounds much more like a *bomb* than a rocket. Gentlemen, let me make this perfectly clear to you. I will not tolerate a bomb club in my school. And as for trophies, Mr. Hickam, your brother and the football team don't need your help."

"But I think these boys have a wonderful idea, Mr. Turner," Miss Riley said. She smiled at me. She had an impish, freckled face. "I graduated from this high school," she said, "and all I ever heard was football, football, football. Wouldn't it be wonderful if science was another way to get in this trophy case?"

"That's just what I was saying, Miss Riley!" Quentin blurted.

"I am disciplining these boys at present, Miss Riley," Mr. Turner said, shooting Quentin a warning look. The bell rang and students started to stream into classrooms up and down the hall. "Well?" Mr. Turner demanded of us. "Don't you have classes?"

"I'm in charge of helping students prepare for the county science fair," Miss Riley said to us over the noise of the throng. "If you boys are interested, come and talk to me."

"Yes, ma'am!" Quentin chirped.

I felt like strangling Quentin. All we had done was blow up a fence and stink up Coalwood with our failures. It was embarrassing. "We can't be in any science fair," I muttered.

Miss Riley studied me. It felt as if she could see right through me. "Why not, Sonny?"

"We just can't," I repeated stubbornly. I didn't want to explain. I just wanted to get off the subject.

"Go away, boys," Mr. Turner waved. "Quickly, now."

I was grateful for the excuse to get away and ran for it. With his big briefcase practically dragging on the floor, Quentin couldn't get anywhere too fast, but he caught me while I waited for the other students to file inside the door to history class. "Listen, Sonny," he gasped, catching his breath, "we win the science fair with our rockets, it's got to help us get on down to the Cape."

Besides the fact we didn't know how to build a rocket, I told him my main objection. "Quentin, we'd just embarrass ourselves. We'd be up against Welch High School students." This was self-explanatory, I thought. Welch students came from families with fathers who were doctors, lawyers, judges, businessmen, and bankers, and their high school was the newest, best-equipped school in the county. The *Welch Daily News* had stories all the time about Welch students going off to college and winning honors. Although we routinely knocked the tar out of them in football, there was no way any Big Creek student was going to beat Welch students head to head in a science fair. "You want it in the paper and everywhere else how we got stomped? How would that look to Dr. von Braun? If you have an ounce of common sense, you'll drop this idea," I told him, perfectly aware that he lacked that ounce.

"It's not like you to be a pessimist," Quentin said coldly. "I'm totally dumbfounded by your attitude. Dismayed too." When I didn't say anything, he added, "Astonished, chagrined, and saddened."

I wasn't going to let him bait me with his vocabulary. I just shook my head and left him standing in the doorway. I didn't want to hear any more about it.

NEARLY every Sunday afternoon that year, I thumbed rides to War to visit Dorothy for study sessions. She seemed to enjoy my company, and it wasn't her fault, after all, that I was in love with her. One Sunday, she stopped studying and looked across the coffee table at me. "Oh, Sonny, I'm so glad we're such good friends!" she gushed.

"Me too, Dorothy," I answered, lying. Never had *friend* been such an awful word.

Emily Sue caught me staring unhappily at Dorothy in the auditorium one morning. Dorothy was holding hands with her latest, a senior basketball player, and I had my lip out about it. Emily Sue sat down in front of me and put her arm up on the seat, looking over it at me. Because she was plump, was a brilliant scholar, and had big, round glasses that gave her face an owllike appearance, it might have been expected that Emily Sue wasn't popular with the boys, but she was. For one thing, she was one of the best dancers in school. But to me, Emily Sue was what I came to think of as a forever friend, somebody I could tell the truth to without fear of reproach. I just instinctively knew that about her. She also seemed to possess a wisdom far beyond our years. "So what are you going to do about her?" she asked me, nodding toward Dorothy.

"Nothing I can do," I shrugged, working hard to be nonchalant.

Emily Sue inspected me. "She likes you, Sonny, but to her you're just her special little friend. That's probably not ever going to change."

Her words were like knives plunged into my heart. I abandoned all pretense. "But why?" I whined. "What's wrong with me?"

"There's nothing wrong with *you*," Emily Sue said. "You're one of the nicest, friendliest kids in this school. Everybody likes you, Sonny. You know why? *You like yourself.* Look at your brother. He dresses great, he's a football star, he's a wonderful dancer—God knows, I love to dance with him—and there's a lot of girls after him all the time. He's a big man on campus, but he doesn't really have any friends. That's why I think he goes out with so many girls. He's trying to find someone who will like him for who he is, not because he's a big football star. Dorothy's the same way. She's happy you're her little friend but she's going to keep looking somewhere else for love."

While Emily Sue was talking, I was sliding deeper into my seat. Jim and Dorothy alike? I wasn't buying that. And me forever just Dorothy's friend and nothing else? The thought of it plunged me into a melancholy as deep as the coal mine. The bell rang and I thanked Emily Sue for the nice things she said about me, didn't argue with her about the rest, and then made my getaway. I didn't want to, but all that day I thought about nothing else but what Emily Sue had said. I just couldn't believe it. There had to be a way to win Dorothy, some strategy, some ploy. It was like building a rocket. I could figure it out. If only I was smart enough.

OVER the next few weekends, Quentin and I continued our practical work, testing different black-powder mixes in the hot-water heater. I was grateful when he didn't bring up the science-fair idea again. Finally, our trial-and-error method resulted in a combination of ingredients that seemed to have the most flash and smoke to it. Quentin had an idea on the propellant. "I've been thinking, old boy," he said. "I don't like this loose mix. It seems to me we ought to put some kind of combustible glue in it so it can be shaped. We could run a hole up through the center of it, get more surface area burning at once. That ought to give us more boost."

A boost sounded good to me. I went to the Big Store and waited until Junior had served the other drugstore customers and then asked him if he knew of anything that was like a glue but would burn. He questioned me closely until I admitted exactly what I was trying to do. He brought out a can of powdered glue. To this day, I still can't imagine why it was stocked at the Big Store, but there it was. "This is the same stuff that's on the back of postage stamps. Mix a little of it in your powder and add water. Let it dry. I think it'll burn. That'll be fifteen cents."

"Thanks, Junior," I said, counting out the change. I was almost out of scrip. I'd soon be cutting into my paltry stash of U.S. dollars.

"I hear you plan on going to work down at Cape Canaveral," he said. "I was in that part of Florida once. Swam in the ocean. Coloreds-only beach."

It had never occurred to me that colored people needed their own beach. I guess it should have, considering in Coalwood they had their own schools and their own church, but it hadn't. "Did you like it?" I asked.

Junior looked uncomfortable. "I liked it just fine. But I took my mother, and she didn't hold with it at all. Said she couldn't wait to get back to the mountains." He pondered a bit. "When she died, we buried her on the mountain behind Little Richard's church."

"Tell Reverend Richard I said hi," I said.

"I don't go to that colored church," he snapped. "I go up on the mountain to pray." He frowned at me. "Go on and build your rocket. But be careful, hear?"

I had upset Junior, but I didn't know how. "I will, sir," I promised.

Junior hurried over to attend to customers lining up at the counter. "I'll tell Little you said hello," he relented just as I went out the door.

Once home, I got out measuring spoons, cups, a mixing bowl, and an egg beater from Mom's cupboard and carried them to the basement. I mixed up what Quentin and I had calculated to be our best black-powder mix and then added the powdered glue to it along with a little water until I had a thick black paste. I wrote down everything that I had done in a notebook. *Body of knowledge.* I poured the slurry in a dish and then set it underneath the water heater to dry. Two days later, it had formed into a hard cake. In search of her missing cups, spoons, and mixing things, Mom came down, looked over my laboratory, sighed, and went down to the Big Store to buy replacements. She told me later she and Junior had a big laugh over it. When Quentin and I threw the black-powder cake in the hot-water heater on the following Saturday, it flashed vigorously. "*Prodigious,* old man!" he cried, using his latest big word for anything he liked.

We kept trying to figure out the "why" of rockets as well as the "how." Although he hadn't found a rocket book, Quentin had finally found the physics book he'd read in the library in Welch that defined Newton's third law of action and reaction. The example given in the book was a balloon that flew around the room when its neck was opened. The air inside the balloon was under pressure, and as it flowed out of the opening (action), the balloon was propelled forward (reaction). A rocket, then, was sort of a hard balloon.

Instinctively, we knew that the nozzle (the opening at the rocket's bottom), like the neck in the balloon, needed to be smaller than the casement. But how much smaller, and how the nozzle worked, and how to build one, we had no idea. All we could do was guess. "How about we weld a washer or something at the bottom of the casement to be our nozzle?" I proposed to Quentin at lunch one day.

Quentin pondered that while chewing on the cookies Mom had sent him in my lunch bag. "Yes. I think that might work. But who would do the welding?"

I knew of three welders in Coalwood. Two of them worked in the big machine shop across from the creek down by the Big Store. Mr. Leon Ferro was their supervisor. I didn't think he'd help us, since he was a company man, like my father. There was, however, a machinist-welder who worked alone up at the tipple shop during the hoot-owl shift. His name was Mr. Isaac Bykovski. Mr. Bykovski's daughter, Esther, had been in my class until she was diagnosed with cerebral palsy and had to go away to a special school. My mom said Mr. and Mrs. Bykovski were always asking about me, how I was doing in school and so on. And when we had school plays, I'd sometimes look out and there they'd be, Ike and Mary Bykovski, smiling at me just like I belonged to them. I thought maybe I had our welder.

That night, after the hoot-owl miners had descended into the mine and the evening shift dispersed, I slipped out the back door and up the tipple path. Just in case Dad was called out by the black phone, I angled away into the trees so he wouldn't see me. I was also heading for a secret entrance to the mine area we kids had discovered years before, playing cowboys and Indians. The tipple area was locked at night, but just inside the tree line on the mountain, there was a deep drainage ditch that went underneath the fence. Nearby was also a locked gate with a seldom-used path that led to the little shop. I found the path and groped my way in the darkness until I came to the gate. I gripped the chain-link fence and edged down the mountain until I stood on the precipice of the ditch. There was a big drainage pipe there that jutted out on the other side. I gripped the links of the fence with one hand and swung down until my feet got a purchase on the pipe, then ducked underneath the fence and came up on the other side. The machine shop was no more than a dozen yards away, its lights illuminating the gate and the path.

I peeked through a dirty glass pane in the back door. Mr. Bykovski, in a baggy one-piece coverall, was working at a lathe. He was a skinny little man with tiny ears that stuck out so far it looked like his helmet was resting on them. Screwing up my courage, I opened the door and walked in. He saw me and nodded, and I waited until he finished what he was doing. "How are you, Sonny?" he asked, as if it were a normal thing for me to appear before him in the middle of the night.

Mr. Bykovski had a trace of an accent, which didn't seem strange to me. There was a sizable immigrant population in Coalwood. Italians had come in to the county as strikebreakers during the 1920's and 1930's and then joined the UMWA during World War II. Hungarians, Russians, and Poles had come in after the war. There were also a couple of Irish and English families and one Mexican. Although the parents in these families spoke with accents, none of their children did. The Great Six made certain of that, our daily lessons filled with the importance of the spoken and written English word. Old-timey West Virginia children fared no better than immigrant children. If we said "far" instead of "fire," or "born-ded" instead of "born," or even "holler" for "hollow," we were instantly corrected and made to say the word over and over again until we understood such pronunciations were not to be tolerated. And only heaven itself could help a Coalwood grade-schooler if he said "liberry" instead of "library."

I told Mr. Bykovski I was building a rocket and needed a washer or something welded to the bottom of a tube. "And you want me to do it?" he asked.

"Will you?" I held my breath.

He took off his helmet and wiped the sweat off his nearly bald head with his sleeve. "I have some aluminum tubing I could use. But welding on a washer—that is difficult. Soldering it would be easier."

"That would be fine," I said. As long as my washer was well attached, soldering sounded good to me, although I wasn't exactly certain what soldering was.

He looked at me sharply. "I am not supposed to do work in this shop unless your father tells me to. Does he know you are up here?"

I shook my head. "No, sir." I had kind of an instinct about Mr. Bykovski. It was best to tell the absolute truth with him, no shading. "He's against me building rockets, but Mom thinks it's okay. I need help, Mr. Bykovski. You're my only hope."

He considered me for a moment, his face grim. I know I must have looked pitiful, because that's the way I felt. "Do you know how to solder?" he said at length.

"No, sir."

"Then I will teach you. Your dad should not have a problem with that. Come on. You can work while I work. How long should your tube be?"

I wasn't certain, so I said a foot. He then asked how wide and I wasn't certain about that, so I said an inch. He cut off a foot of one-inch-wide aluminum tubing, and it looked awfully small. He went for a larger diameter, an inch and a quarter, and that looked better. He also increased the length to fourteen inches, and that looked about right to me. He then gave me a quick lesson in soldering. It seemed simple enough. All you had to do was hold a hot iron rod to a coil of solder, which was like a soft metal, melt it, and let the silvery material flow into place. It proved harder than it looked. I gobbed on the melted liquid, but I made a mess of it, the solder dripping down the tubing and the washer not on straight. After an hour, Mr. Bykovski came over to see how I was doing. "It is not bad for your first time," he lied. "I will finish it up for you during my break. Come back tomorrow night and I will have it ready."

He didn't have to make that suggestion twice. I was almost ready to pass out from the lack of sleep. The next night I again went up my secret path and found my rocket waiting for me in a cardboard box outside the gate. The solder was a perfect circle around a perfectly aligned washer at the base, and he had also soldered a metal cap to the top of the tube and glued on a wooden bullet-shaped nose cone. It was the most beautiful rocket I'd ever seen. I used electrical tape to attach cardboard fins to the casement and then borrowed Mom's fingernail polish to paint a name on the side. I named it *Auk I,* after the great auk, an extinct bird that couldn't fly. Quentin, for no apparent reason, had gone on and on the day before about extinct birds, so I was up on them. I had a purpose in the name. I wanted to make it clear to the other boys that we were adding to our body of knowledge even if all this rocket did was spew on its launchpad.

I loaded *Auk I* with the black-powder/postage-stamp-glue slurry mix, inserted a pencil through the nozzle, and then left the rocket to dry under the hot-water heater. The pencil was to form a hole in the powder, increasing its surface area, according to Quentin's idea.

On Saturday, Quentin hitchhiked to my house, and after the other boys arrived, we inspected *Auk I.* "Love the name," Quentin said. "Maybe the gods will help us, thinking we are suitably respectful of pernicious fate."

The other boys looked at him blankly. "It's bad luck to be overconfident," I translated.

"I believe we're making progress," Quentin continued, talking to me as if the other boys didn't exist. He ran his fingers around the base of the rocket, studied carefully the soldered washer, sniffed speculatively at the cured black-powder mix. Then he said, "But we cannot simply proceed by trial and error. This rocket may fly. If not, then the next one shall. But what will we have learned, seeing this tube shoot up in the sky like nothing more than a Fourth of July skyrocket? That's not what this is about. We have to learn *why* it flies."

"That's your job, Quentin," I snapped angrily. I thought I had done everything he wanted, and here he was criticizing. "When are you going to find us a book?"

He shook his head. "I don't know where else to look. Maybe it's so secret they don't write it down."

The other boys shifted restlessly. "Can we just go launch the damn thing?" O'Dell demanded.

"O'Dell," Quentin replied, in all sincerity, "I'm worried that your insatiable cupidity will ultimately prove to be something less than a virtue for our club."

O'Dell stirred from his seat menacingly. "How about I use my insatiable cupidity to beat the crap out of you?"

I knew I'd better head off trouble. I didn't mind when Quentin showed off his vocabulary, but I suspected the other boys thought he was just being obnoxious, which, of course, he was. "Let's go. Sherman, lead the way. O'Dell, you carry the rocket. Roy Lee, you got the matches? Quentin, you stay with me."

Sherman led us up Water Tank Mountain to the old slack dump. We were at least two hundred yards above the mine. I could just make out the top of the tipple above a stand of trees. O'Dell sat the rocket on its base and then used a rock to hold it steady. We all found hiding places behind big boulders around the clearing. O'Dell took a match from Roy Lee. "A rocket won't fly unless somebody lights the fuse!" he declared. Sherman settled in behind a rock. O'Dell lit the fuse and ran and fell down beside me. We grinned at one another.

The fuse sizzled up inside, and *Auk I* leapt into the air in a shower of sparks. Six feet off the slack, it made a *poot* sound and then fell back in a cloud of gray smoke and landed heavily, breaking off its nose cone. There it lay until the powder quit burning. Quentin got to it first, getting down on his hands and knees and peering at the rocket's base. "The solder melted," he announced, wrinkling his nose at the sulfurous stench. "It was flying, but the solder melted."

When it cooled, I picked up the aluminum tube. It stunk, but it had flown. It had gotten only six feet off the ground. *But it had flown!*

"Prodigious," Quentin said.

On Sunday night, I once more went under the fence to see Mr. Bykovski, carrying *Auk I* with me. He examined it. "Rockets get too hot for solder, looks like. It's going to take a weld after all." He pushed his helmet back on his head, a cogitative move. "It is a hard thing to weld aluminum. Steel would be better."

He went over to his racks of materials, selected a steel tube, and cut off fourteen inches of it with a hacksaw. He handed it to me and I hefted it. "Feels heavy," I said dubiously.

"Yes, but steel is strong, Sonny," he said. "An aluminum tube, to be as strong, requires a very thick wall. With steel, the wall of the tube can be thinner. I recommend it to you. I have been thinking, also, about the washer. That is not a good metal. I think we must cut off a thin piece of steel-bar stock, drill a hole in it, and weld it to the base."

I absorbed all that he had said. "Will you teach me how to cut steel bars and drill them and how to weld?"

Mr. Bykovski looked at his watch. "It would be quicker if this time I do it for you. I will teach you how to do it yourself another time."

My conscience was pricked, just a little. "I wouldn't want to get you into any trouble," I said.

He shrugged. "Your father will not find a better machinist than me, especially one who will work the hoot-owl shift. I think, anyway, you should tell your father. He ought to be proud of what we do here."

"When our rocket flies—really flies—I'll tell him then," I promised halfheartedly.

He beamed. "Good. This next one will fly, I think. I will have it ready by this Wednesday."

I decided to push my luck. "Mr. Bykovski, could you make me two?"

He made me three. On the following Saturday, *Auks II, III,* and *IV* were ready, built exactly as he had described. Once more we went up to the clearing behind the mine. "A rocket won't fly unless somebody lights the fuse!" O'Dell said, explaining since he'd said it when our first rocket flew, he thought maybe it was a good-luck thing to keep saying it.

Sherman wanted to light the fuse, but I worried whether he could get away from it fast enough. "Don't you worry about me," he said with such intensity that I instantly gave in. In a lot of ways, Sherman was the least handicapped person I'd ever known. He lit the fuse and ran back to a rock. Flames burst from *Auk II.* It sat for a moment, spewing smoke and sparks and rocking on its fins. Then it jumped ten feet into the air, turned and zipped into the woods behind us, ricocheted off an oak tree, rebounded back to the slack, twisted around once, twanged into the boulder Quentin and I were hiding behind, jerked twenty feet into the air, coughed once, and dropped like a dead bird. I had kept my eye on it the whole way, but Quentin had more wisely buried his face in the slack, his hands over his head. I tapped him on his shoulder and he jerked his head up, slack spilling out of his nose. "It's down," I told him as I crawled to my feet. O'Dell ran to *Auk II* and began a wild little dance over it. "It flew! It flew!" he sang.

"It almost killed us," Roy Lee said hoarsely as he climbed out of the ditch he'd thrown himself into. He walked up to O'Dell and waited patiently until the boy had stopped gyrating. He kicked at the hot rocket. "But it *did* fly, didn't it?"

We were all shaking with delight. "I lit the fuse!" Sherman crowed.

Quentin brushed himself off and inspected *Auk II*. His nose was smudged black. "We need to come up with a better guidance system before launching again," he said.

The rest of us weren't having any of it. Our rocket had flown! We ached to see what the next one would do. This time, Roy Lee lit the fuse and tripped, cursing. He barely made it to a boulder before the rocket blasted off, twirled around once, twanged off a maple tree, bounced off the ground near us, and then thudded into the side of the mountain above us, nearly burying itself into the dirt.

While the rest of us joined O'Dell in another celebratory dance, Quentin dug out *Auk III*. "I'm telling you we'd better not launch again until we figure out how to make these things go straight," he said.

Roy Lee gleefully set *Auk IV* up. "We came up here to fly these rockets, and that's what we're going to do." Without further ado, he lit the fuse. Caught unawares, the rest of us had to scramble to get behind our rocks before the fuse reached the powder.

With a *whoosh, Auk IV* climbed smoothly into the air and headed down the mountain. I raised a cheer that turned into a strangled yelp when I realized the rocket was heading for the mine. I had a momentary vision of our rocket falling down the shaft like a torch being dropped into a deep tank of gasoline. When I saw that the smoke trail was angling off to the left of the tipple, I knew we'd escaped at least that particular disaster. Still, I had little doubt there was going to be trouble. I felt a deep heaviness in my stomach, like all my insides were going to come out through my toes. I felt like kicking myself for launching so close to the mine. I was the leader of the BCMA, and it was my fault. *How could I have been so stupid?* I knew the answer, couldn't blame it on anybody but myself. I'd agreed to launch it close to the mine because I didn't really think our rockets were going to work.

There was no sense in all of us going after our rocket, so after a brief discussion, it was decided Quentin and I would go. He and I, after all, had designed what had suddenly become to O'Dell "the blamed thing." The others headed off to circle around the mine to the road. I steeled myself for what was going to happen next, hoping that maybe, just maybe, *Auk IV* had fallen somewhere where we could retrieve it without being seen. We silently stole down the mountain and slipped like ambushing Indians through the open gate in the back of the tipple area, where Dad's grimy little brick office was. It was Saturday, but Dad was at his office as he often was on the weekend. He saw us before we saw him. The miners had a special yell, a "whoop," when they wanted to get attention in the mine, and they often used it on the surface too. I heard Dad's whoop and looked up and saw him on his office porch with two other men. They were dressed in coats and ties. They had to be men from the Ohio steel mill that owned us, since they were the only men I ever saw wearing a coat and tie in Coalwood except at church or the Club House when there was a party. I spotted *Auk IV* lying forlornly in the coal dust beside the railroad track. There was a big chunk out of the brick wall of Dad's office. It didn't take a genius to realize what had happened, or how it would be perceived. The BCMA had rocket-attacked the coal company.

In the darkness of the mine, the signal from one miner to another to approach was a roll of the head so the light from his helmet lamp made a circle. They were so used to doing it, I'd seen helmetless miners, forgetting they were outside in the full daylight, roll their heads when they wanted somebody to come in their direction. Dad rolled his head at me, and I got to him in a hurry. He was so mad he was huffing and puffing. I was afraid he was about to go off into another one of his coughing fits. "I told you to stop this, didn't I?" he barked. "You could have killed somebody with that thing!"

I was at least relieved to hear our rocket hadn't hit anybody. Dad came off the porch and picked the rocket up. "This looks like company property to me. Where did you get it?"

I was too scared to reply. It wasn't that I was afraid of being hit or anything like that. My father had hit me only once in my entire life. I was about seven years old and playing with my dog Littlebit around the old mine shaft down from the tipple. I climbed inside the old shack that enclosed it, just to look down the deep, dark hole. Littlebit came inside and ran to me. I don't think he even noticed the hole until he was almost on it. He leapt for me and almost made it. It was a six-hundred-forty-foot drop, straight down. Dad brought Littlebit's limp body home that night and, while I wailed, turned me over his knee and gave me three good whacks. Then he helped me bury Littlebit on the other side of the tracks. I made him take his helmet off so I could say a prayer. "Dear God," I whimpered inconsolably. "Please kill me too, because I got Littlebit killed."

"That's a terrible prayer!" Dad gasped. "Do it over. Pray for Littlebit's soul or something."

"All right," I said dutifully. "Dear God, please let Littlebit be happy in heaven, and please don't kill me even though I deserve it."

"Gawdalmighty, what do they teach you in that church?" My father only made rare appearances at the Coalwood Community Church. He put his big hand on my shoulder. "God, he's a child. Bless him"—he hesitated—"as best You can." Then he plopped his helmet back on. "Come on. Let's go see what your mom's got for supper."

Now one of the Ohio men laughed and the other one joined in, their laughter like braying mules. "Looks like your boy wants to be a rocket scientist, Homer!"

"He doesn't know what he wants to be," Dad said, leveling a steely gaze at me. "But I know what he is." He held up the rocket. "He's a thief." He inspected the weld at the base. "And so is the man at this mine who helped him."

7

CAPE COALWOOD

DAD WASN'T INTERESTED in him, so Quentin escaped from Coalwood, hitching a ride at the mine entrance. The other boys, I assumed, had gone home to hunker down for the day, hoping without much hope that their parents wouldn't hear about our errant rocket. Dad ordered me to walk home. He followed about an hour later and called me out into the yard. I waited while he went down in the basement and returned with my chemicals in a cardboard box. "Come with me," he said. "I want you to see this." I followed him out the back gate and then watched as he poured everything into the creek. I knew he was justifiably angry, considering how stupid I'd been to launch our uncontrolled missiles so close to the mine. On the other hand, these were *my* chemicals paid for with *my* money. I'd gotten up on a lot of cold, snowy mornings to deliver the paper and earn that money. "This is the end of it," he said over his shoulder as he shook out the last bag of saltpeter, "and this time I mean it. Collect stamps, catch frogs, keep bugs in a jar, do whatever you want. But no more rockets." He handed me the box filled with empty bottles and bags. "Now, who helped you?"

I remained silent, but he said, "Bykovski. Got to be." I felt my face involuntarily slide into an expression of dismay. Was there anything in Coalwood my dad didn't know about? "I'll take care of *him*," he assured me.

"What are you going to do?" I asked urgently.

"That's none of your business, little man. Now go up to your room and stay there until your mother gets home."

When Mom got home, Dad stopped her at the door. I could hear them talking, but not exactly what was being said. Then I heard her thump up the stairs. She came into my room. "Tell me what happened," she said wearily.

I gave her the whole story, about Mr. Bykovski and everything. "I wondered where you were sneaking off to in the night," she said after I finished. "Don't look so surprised. You think I don't know what goes on in this house?"

"Are you going to help me?"

She shook her head. "I don't see how I can. The Ohio men told Mr. Van Dyke what happened. Your dad's pretty embarrassed, and I guess he has a right to be."

"What should I do?"

"I don't know. You messed up pretty good this time."

"I guess I'm finished," I said.

"If you give up that easy," she replied with a shrug, "I guess you are."

"I'm worried about Mr. Bykovski," I said, looking for sympathy.

"You should be," she answered coldly. "You used him. Ike and Mary have always had a special liking for you, and you knew it. You should have thought of what could happen to him before you got him involved."

I sweated out the rest of the day and then, as soon as the shift change was made, slunk up to see Mr. Bykovski. I was relieved to see him in his shop. He was working on the big steel-cutting maw of a continuous mining machine. He saw me at the door and waved me inside. "You see, Sonny?" he said, pointing at the maw. "The operator hit rock instead of coal. The teeth have been broken off. I will build new ones."

I picked up one of the broken teeth on his worktable and fingered it. "Did—did my dad talk to you?"

"Your father was pretty mad," he said over the shriek of a milling machine. "This is my last night in the machine shop. He reassigned me to the mine. I'll be operating a loader on the evening shift."

Revulsion and shame welled up inside me. I had acted stupidly, but Dad's reaction was vile and despicable. "My dad's the meanest man in this town!" I erupted angrily.

Mr. Bykovski stopped the milling machine and came over and grabbed me by my shoulder, giving me a good shake. "You must not say this about your father. He is a good man. I acted without his permission, and I deserve to be punished." He released me and patted me on the side of my arm. He smiled a sad smile. "Anyway, perhaps it is a good thing he has done. I will make more money loading coal."

"I'm sorry, Mr. Bykovski," I said. "Mom said I took advantage of you, and she's right."

"Look, I have something for you," he said. He went to the tool crib and carried out a cardboard box. Inside were four new *Auks,* complete with wooden nose cones. "I already had them made up. They should keep you going for a while. Now, go on. I have much work to do."

I clutched the box as if it were filled with gold and diamonds. "I'll never be able to thank you enough."

"You want to thank me?" He nodded toward the box. "Make these fly. Show your dad what you and I did together."

My father had clearly, in no uncertain terms, told me to stop building rockets. The BCMA was now an outlaw organization. I don't know why, but that felt good. I had the urge to hug Mr. Bykovski, but resisted it. Instead, I stood straight and tall and said firmly, and what I hoped was manfully, "Yes, sir. You can count on me."

He nodded and went back to work. So did I.

O N the following Monday, I gathered the boys in the Big Creek auditorium before morning classes. As expected, the gossip fence had instantly informed their parents about our assault on the tipple. Surprisingly, all of the other boys had gotten off without punishment. Roy Lee's mother had laughed it off. O'Dell's father thought it was pretty amazing the rockets had flown at all and, after all, no harm had been done. Sherman's father had counseled him to think about things a little more before he did them, but that was all. I was the only one who'd been yelled at. When I reflected on it, I suspected the other parents thought it was funny that we had spooked the Ohio men, who were not exactly beloved by the average Coalwoodian. I'd heard Roy Lee, who got the union talk from his brother, say the steel mill muckety-mucks were far more interested in themselves than us, that they'd sell us down the river in a second. My father, on the other hand, believed a major part of his job was keeping the men from Ohio happy. Well, I had myself to keep happy. "We've got to get a new rocket range, someplace out of Coalwood," I told the boys.

"You mean we're not quitting?" O'Dell asked.

"We're outlaws now," I said, savoring the word. "We're not ever going to quit."

Sherman was with me. "They clear-cut all the timber off Pine Knob," he said. "It's not on company property. We could go up there."

"Are you kidding?" Roy Lee griped. "We'd have to climb two mountains to get up there."

"Do you have a better plan?" Sherman countered.

"I sure do. How about we stop all this rocket stuff and get us some girlfriends?"

That interested O'Dell. "How do we do that?"

"First I'd need to teach you the ropes."

"Like what?"

Roy Lee's eyebrows went up and down. "Like unsnapping a bra with one hand."

"Pine Knob it is," I decided, ignoring Roy Lee's nonsense. "This Saturday, meet at my house. We'll leave from there. Quentin?"

"Yo," Quentin answered, shaken from some distant reverie.

"We need to figure out a better way to test our mixes other than just throwing them into the hot-water heater. You're supposed to be our scientist. Can you think of some way to do that?"

"Of course."

"Do it."

On Saturday, while Dad was at the mine, the BCMA met in my

room. Quentin had labored over a way to test our powder all week and proudly presented his plan. It was a complicated test stand with tubes and springs and pistons. I was impressed. It looked like something Wernher von Braun himself might have dreamed up. O'Dell was the first to speak after Quentin's breathless explanation of how it all worked. "How about we just put the powder in a pop bottle and see how big an explosion it makes?"

A chorus of agreement followed. Everybody looked at me to make the decision. "Pop bottle," I decided. I hated to disappoint Quentin, but we had no way to build his design with our limited resources. "But good job anyway, Q," I said. I had already figured out it never hurt to give somebody a pat on the back.

Quentin protested. "Sonny, we must approach this enterprise in a scientific manner!"

"We are, Quentin," I told him calmly, "but we have to sometimes take into account we're not at Cape Canaveral."

Quentin appealed to the other boys lounging around the room. "We're trying to learn how to build a rocket, gentlemen. This isn't for fun."

"How right you are, Quentin," Roy Lee said, winking at me. "That's what girls are for." From his jacket pocket, he brought out a brassiere and wrapped it around a chair. "Okay, as I promised you, it is time to watch and learn, boys."

Quentin sighed in exasperation. I crowded in with Sherman and O'Dell. We were eager to learn Roy Lee's adult secrets. Roy Lee sat down in a chair beside the one with the bra and draped his arm around its top. After a moment of deftly fingering the attachments in the back, the bra fell apart. "Wow," we all said in unison, even Quentin. He picked up the bra and inspected the complicated hooks and loops in the back. "You know," he cogitated, his brow furrowing, "there should be a better system." He plucked a beggar's lice—endemic in West Virginia—off his pants leg and inspected the tiny fuzzy seed that hitched rides on anything or anybody who walked through the woods. Dandy and Poteet used to come back from chasing rabbits covered with them, and I'd spend hours picking them off. Quentin put the seed back on his pants and then pulled it off again. "I'd like to look at this under a microscope. If you could figure out what makes it stick to your pants, you could maybe put it on cloth straps and—"

"Shut up, Quentin," Roy Lee said, snatching the bra from him and strapping it to the chair again. "You think too much."

One by one, we each took our turn at the thing. I'd seen lots of brassieres hanging outside on washlines up and down Coalwood, but had never had occasion to touch one before. Unhooking it one-handed wasn't nearly as easy as Roy Lee had made it look. The top hook was the hardest. "Dorothy would have slapped you silly by now," Roy Lee told me.

"Don't talk about Dorothy that way," I bristled.

"Why? She's no angel. I heard she's dating some boy over in Welch."

That was news to me. The kids in Welch were considered "fast" by the rest of us in the county. If she was going out with a boy from there . . . I felt butterflies in my stomach. "Just let it be, Roy Lee," I snapped, suddenly miserable. It seemed everything about Dorothy either made me very happy or very sad.

Roy Lee gave me his best innocent look and held up his hands. "Okay. But don't say you didn't know about it."

We kept up our bra work for the rest of the afternoon until we all had it down to a science, even Quentin, who finally gave in to Roy Lee's vivid descriptions of what might occur if you could master such a handy talent. After the other boys filed out of the house, Roy Lee with the bra hidden back in his jacket, Mom stopped Quentin and asked him to stay for supper. He gave her his little bow. "I'd be delighted, Mrs. Hickam, just for the pleasure of your company."

She grinned with delight. "Sonny, why don't you have manners like Quentin?"

"My upbringing?" I asked.

"A smart mouth could get a boy in trouble," she warned. "You've really got the basement in a mess. You want to clean it up down there?"

"Yes, ma'am," I said, and reminded myself that my mother was never a person to cross, even a little.

Over the next week, O'Dell gathered the pop bottles. There were always plenty in the garbage. Quentin and I mixed up a variety of black-powder samples. We began our excursion up to Pine Knob the following Saturday, each of us carrying a paper bag full of bottles filled with different mixes. We had to first climb Water Tank Mountain, so-called because of the two cylindrical steel water tanks on top that held Coalwood's drinking water. Once we reached the tanks, we had to go down the backside of it and then climb up a gully until we reached Pine Knob's bald top. Somebody had clear-cut the top of the mountain more than a decade ago, and the forest still hadn't recovered. There was just a sea of ugly stumps and barren, eroded dirt.

After Quentin got through complaining how much his feet hurt and how tired he was, we set up and began our tests, blowing the bottles to smithereens, one by one, while we hid behind the stumps. Sherman took notes. Quentin kept grumbling about our unscientific approach. In truth, the tests were subjective at best, one exploding bottle difficult to compare to another. I had ground the powder especially fine for the last bottle we detonated, however, and when it went off, it blew a crater a foot deep. Even Quentin was impressed. "That's the way I'll mix it and grind it for our next rocket," I told him, and he was mollified. We had accomplished something nearly scientific, after all. The finer the powder, the bigger the explosion.

Despite the fact we were off company property, we would later hear that some people in Coalwood were still unhappy over what we were doing. It seemed like every time we blew a bottle, some people thought the mine had blown up. They'd come out on their porches in a panic, but then somebody else would say, no, *it's just those damn rocket boys!* Then everybody would go back inside until we blew another bottle, and the same thing would happen all over again. When I returned, Mom said Dad had gotten phone calls about the noise, including one from Mr. Van Dyke. Surprisingly, Dad didn't say anything to me about it. I was off company property, so perhaps he felt I had followed his orders, at least to an extent that satisfied him. But I still had a problem. Pine Knob was okay to blow up bottles, but it was impractical for a rocket range.

The BCMA needed, more than anything else, a place of its own nearby where we could launch our rockets without anybody complaining. But where could we find such a place? As it would turn out, that particular decision was about to be taken out of my hands.

AT the Coalwood Women's Club meeting on Thursday evening, the Great Six teachers met with Mom and Mrs. Van Dyke and gave them their unsolicited advice on the whole matter of us outlaw rocket boys. Mom woke me up the next morning. It was still dark outside. "Come on. We're going to talk to your father."

Bleary-eyed and confused, I followed her to the kitchen. Dad almost dropped his coffee mug when we appeared. It was a rare thing for Mom to be up that early, and he had never seen me before sunrise, ever. "Damn, Elsie, don't sneak up on me like that!"

"Sonny and I need to talk to you, Homer," Mom said.

Dad eyed me and sagged in his chair. "Let's hear it."

"I want you to figure out how he can launch his rockets without everybody getting upset about it."

"Why would I want to do that?"

"Because there's some people in this town who think he and the other boys are trying to do a good thing."

Although I didn't have a clue, Dad, of course, already knew who Mom was talking about; don't ask me how. "Those damn old biddy schoolteachers think all they have to do is snap their fingers and they get their way about everything in this town." He gulped down the dregs of his coffee. "Sorry. This comes from higher than me anyway. Van Dyke says no more rockets too."

"You'll come to regret this, Homer," Mom said icily. She wrapped her housecoat around her and stalked out of the kitchen.

I was left alone, standing at attention before Dad, who finally acknowledged my presence. "Do you see all the trouble you've caused?" he demanded.

I was confused about exactly what trouble he had in mind. After all, I hadn't fired any rockets since his last order to stop. I did, however, have his complete attention, a rare occurrence. I took advantage of it. "Dad, is Coalwood going to be torn down?"

He looked at me as if I were out of my mind. "What are you talking about?"

"I heard the easy coal's almost gone and the company's going to pull out."

He turned his head as if to face the mine, but saw Mom's beach picture instead. He puzzled over it a moment, as if it were the first time he'd ever seen it, and then turned back to me. "There's more than fifty years of good coal left in that mine."

"That's not what Mr. Dubonnet said."

Dad gripped the edge of the table as if he were going to get up and come across the kitchen after me. He subsided. "Dubonnet's a union rabble-rouser. I don't want you talking to him anymore. I'm a company man, and that makes you a company boy, understand?"

I understood more than he knew. After all, I was the one who had been beaten up by the older boys when the union went out on strike. He didn't know that. I was irritated enough to tell him, but then the black phone rang. He rushed past me into the foyer, snatched up the receiver, and yelled in it before whoever was on the other end had time to say a word. "I'm coming, damn you!"

ON Sunday, Mom and Jim and I got up as usual and dressed to go to church. Dad also came down the steps, decked out in a suit and tie. Mom could not have been more astonished if he had appeared naked. It turned out Mr. Van Dyke had asked him to attend a "brunch" at the Club House after church services. "Well, la-te-da," Mom sang. "A buh-runch. Aren't we fancy?"

Dad frowned at her. "Elsie, I told him we'd both be there."

Mom pushed me ahead, straightening my tie with one hand around my neck. "I'll think about it," she said, and I could almost hear Dad's teeth grind.

Mom and Dad sat in the back of the church between the Van Dykes and Doc and Mrs. Lassiter. Jim sat with the football boys, all of them still grumpy about being left out of the state-championship game. Actually, Dad had mollified my brother, if only a little. Jim now had exclusive use of the Buick every Saturday night. He didn't even have to wash it first if he didn't want to. I wasn't jealous of his privilege. I had only just turned fifteen and didn't have a driver's license yet, and, anyway, Jim had girls to take out. I didn't, and at the rate I was going, it didn't look like I ever would.

I found Sherman and O'Dell and we sat up front. When the choir got up to sing, Mrs. Dantzler stepped out for a solo. She stood straight and tall in her maroon robe, and when the sun came through the clear glass windows, her hair glowed like molten silver. When she finished, her voice seemed to still be ringing up in the rafters. The Reverend Lanier stood up and made his way to the pulpit. I thought something about him didn't look right. He looked a little harried. His robe seemed to be ill-fitting, and his hair needed combing. "Today," he began in a strangely nervous voice, "I will speak on the general topic of fathers and sons."

We live, he said to the hushed congregation, in a day and age when fathers often do not receive the respect they should from their sons. At this remark, my antennae went up. The company paid the good Reverend's salary and was not above suggesting the topic of his sermons, mostly having to do with rendering to Caesar what was Caesar's. What sons might not be respecting their fathers? Who else but us outlaw rocket boys?

Reverend Lanier told us a little story. There was once a son who did bad things, and every time he did one of them, his grieving father drove a nail in a door. When finally the son came to his senses, his father forgave him and removed each of the nails.

"But though the nails were gone," Reverend Lanier said sadly, "the holes were still there, representing the pain still abiding in the father's heart."

Involuntarily, I slid down in the pew when Reverend Lanier looked directly at me. He had worked a little magic, making me feel guilty for something I didn't really feel guilty about. Preachers seemed to be good at that. He talked a little more about the poor, abused door and what it meant and then ticked off an appropriate proverb. Just in case I still doubted who he meant it for, he locked his gaze on me once again. *A foolish son is the calamity of his father. Cease, my son, to hear the instruction that causeth to err from the words of knowledge.*

I slid even further down in my seat. I could just imagine my father smiling smugly at my mother. I thought I knew the real reason my dad was at church services. The preacher was delivering a company sermon!

But the Reverend wasn't finished. He took a deep, nervous breath. This time he didn't look at me but over my head, toward the back pews. "I was taught the story of the door in Bible college, and it's always stuck with me. I've used it more than once to counsel young men who've gotten a little too rambunctious. But the events of late right here in Coalwood have gotten me to think-ing: What of the father who drove those nails in the door? Rather than hammering nails with anger, what if he had instead gone to his son and shown his love with his time, his interest, his generos-ity? Perhaps the holes in the door are a reflection of the father's petulance more than his love."

After clearing his throat and tugging at his collar, the Reverend continued. "We've had some trouble in Coalwood lately," he said, his voice nearly cracking. "Father–son trouble. Of course, the son must respect his father. But I am also reminded of Proverbs twenty-three, verse twenty-four. *He that begetteth a wise child shall have joy of him.* To have a child who longs to learn is the sweetest gift of all."

Although usually there was only silence in the congregation, I heard a few hearty "amens" from the choir box, and then, to my joy, I realized who had influenced at least the second part of this particular sermon. It wasn't the company. It was the *old biddies* who filled most of the choir box. The Great Six. Reverend Lanier wasn't preaching to the choir, he was preaching *for* it.

The Reverend was on a roll. "Sons, obey your fathers. But fa-thers, help your sons to dream. If they are confused, counsel them. If they stray, search them out and bring them home. Our Lord said: If a man has a hundred sheep and one of them strays, does he not leave the ninety-nine on the mountains to go out in search of the stray one? And if he manages to find it, I assure you that he is happier over that one than the ninety-nine that did not stray. Fathers, I beseech you to seek out your straying sons and rescue them by keeping their dreams alive. These boys, and we all know I'm talking about our very own *rocket boys,* are dreaming great dreams. They should be helped, not stifled."

"Amen," the Great Six rumbled as one.

Sherman and O'Dell and I looked at one another and grinned. I heard furious whispering among the football boys. But behind us, where our parents sat, there was nothing but stone cold silence. Reverend Lanier looked intently there and then wiped the sweat off his forehead with the blousy sleeve of his robe. A moment before, he seemed to be lifted high above the congregation, carried aloft by his own rhetoric. Now, perhaps by the reaction he was receiving from certain individuals behind me, he was back on earth. "Of course, this is just a poor preacher's opinion," he said in a shaky voice. His eyes darted. "Ummm, the choir will sing now."

Reverend Lanier sat down, hiding behind the pulpit, but the choir stood up and sang "Faith of our Fathers" with a special enthusiasm. Afterward, the Reverend was supposed to get up and give the blessing, but he stayed seated. After a moment of hesitation, Mr. Dantzler, who had been elected the company-church deacon for the year, rose and asked for everybody to wait until the choir filed outside. I turned around and saw Dad and Mr. Van Dyke looking sour. Mom and Mrs. Van Dyke had angelic smiles on their faces. As the Great Six filed by, they gave us rocket boys stern looks. I knew what that meant. They'd gone out on a limb for us. We'd better do good.

Jim and most of the other football boys went off down the street, but Buck caught Sherman and O'Dell and me out on the church steps. "The Reverend and everybody else ought to be worrying about the football team, not you sister morons."

"Izzat so?" O'Dell made as if to take off his jacket. "Come on, big boy, I'll show you who's a sister."

Mom came out of the church at that moment and saw what was afoot. "Hello, Buchanan," she said to Buck.

"Hullo, Mrs. Hickam," he said, straightening his tie and his posture all at once. "How are you today?"

"Fine, Buchanan. And you?"

"Jus' fine," the big boy said. He looked at us threateningly and then trudged after the other football boys, who were gathering on the Big Store steps.

O'Dell and Sherman both took off in the opposite direction, leaving me with Mom. She said, "Why don't you wait here a minute, Sonny? I think your dad would like to talk to you." Then she walked with Mrs. Van Dyke toward the Club House. I looked over at the parking area and saw Dad and Mr. Van Dyke surrounded by Coalwood teachers. When they broke free, the two men huddled and then Dad sought me out and tossed me the keys to the Buick. "Let's let you practice," he said, looking and sounding thoroughly disgusted.

"Really?" I was delighted at the prospect. Usually, I had to beg him for days to get in a practice session.

He slumped into the passenger seat. "Go toward Frog Level."

The drive was quiet, with me carefully steering while Dad said little except to grumble when I accidentally hit a pothole. When we reached the Frog Level camp, he nodded toward the dirt road, which led down through the wilderness area that was called Big Branch. I took it and was even more careful on the rutted clay lest the big Buick lose its oil pan. Two miles farther down the road, he told me to stop at an abandoned slack dump. "Let's take a look," he said.

We walked out on the gritty surface of the dump. Bulldozers had flattened millions of tons of coal tailings to create a black desert that stretched far down the narrow valley. No tree, not even a blade of grass grew on it. "If you want to fire off your rockets, here's the place," Dad said. "Nobody in town can see or hear you. You've got the entire valley."

I gaped at the huge, flat black space. "How long is it?"

"About a mile, more or less."

I peered down the sun-baked dump and then at the surrounding mountains, my imagination clicking into overdrive. I could see everything as it was going to be: the blockhouse, the launchpad, and our rockets blasting off, roaring up between the steep hills, falling downrange . . . "Cape Coalwood," I breathed.

Dad looked around at the barren slack and then shook his head. "If you want this place, you're the only one. Let's go."

"Dad, there's just one thing."

"What?"

I felt reckless. "We need a building—a blockhouse, it's called—where we can go for protection when our rockets are launched. Could you give us some lumber to build one?"

Dad took off his gentleman's fedora and tapped it impatiently on his leg. "Company property is for company business, not for launching rockets."

"Scrap lumber will do," I explained, sensing that this was my moment. "And can we have some tin for the roof?"

Dad plodded back to the car and then turned and pointed at me. "If I get you this scrap—and even scrap is expensive, young man—if I do this thing, from here on I want the rocket-launching business out of sight and out of mind in Coalwood. Understood?"

"Yessir. Thank you, sir."

Dad perched his fedora back on his head. He looked relieved, but then anxiety played across his face. "Let's go," he said urgently. "God only knows what stories your mother is telling the Van Dykes."

I looked back over my shoulder at the vast slack dump as I followed my father back to the car. The BCMA finally had a home. Cape Coalwood. I couldn't wait to tell Quentin.

8

CONSTRUCTION
OF THE CAPE

AT THE OTHER cape, the one in Florida, business was booming. The Air Force was launching ballistic missiles every week. Most of them blew up, spectacularly, but a few wobbled downrange. On February 5, 1958, the hapless Vanguard team tried again for orbit and failed, although this time their rocket managed to at least clear the gantry before it blew up. On March 17, they gave it another shot, and this time orbited a 3.24-pound satellite nicknamed *Grapefruit*. Dr. von Braun launched another thirty-one-pound *Explorer* into orbit on March 26. It seemed the United States was on the move. Then, in May, the Soviet Union orbited *Sputnik III*, weighing in at a whopping 2,925 pounds. Some Americans, the same kind I thought would have deserted at Valley Forge or surrendered after Pearl Harbor, said we might just as well give up on space.

Dr. von Braun wasn't giving up, not by a long shot. According to a newspaper report, he was building a huge monster rocket called the *Saturn*. In the spring of 1958, Congress and the Eisenhower Administration set up the National Aeronautics and Space Administration in an attempt to put some order into the space program. I read where Dr. von Braun said he might leave the Army and join NASA. If he did, I knew the new agency was my ultimate goal as well.

As my year in the tenth grade at Big Creek dwindled to days, Mr. Turner held one of his few command performances in the school auditorium. We all expected a lecture on school spirit, how we should think about the football team all summer and be ready to cheer them on come fall. Roy Lee and I sat down together. I felt a punch on my shoulder and looked around and saw Valentine Carmina grinning back at me. Valentine, in the class ahead of me, had a figure that was usually described by boys watching her sashay by as "stacked like a brick privy." "Hi, Sonny," she said, her lips parting to show her fine white teeth.

For some reason, Valentine had always liked me. If the other boys were off doing something else and she saw an empty chair beside me in the auditorium in the morning, she'd sometimes ease into it and just talk about things. She came out of Berwind, which was one of the grittiest and roughest towns in the county. She was the oldest of seven children and, she said, had raised all the other kids because her mother was "worn out." She'd also had her problems with Mr. Turner. The faction she led in the Sub-Deb girls' club had been admonished to stop wearing low-cut dresses, smoking in the rest rooms, and sneaking out of class to smooch with boys in the band-instrument storeroom. She had responded by showing up with her dresses hemmed from the bottom and scalloped from the top, threatening, as Mr. Turner lectured her, "to meet somewhere in the middle." Valentine gave in on the clothes, at least by keeping her Sub-Deb jacket on, however artfully unzipped. She could stop all masculine traffic down the hall, with boys' knees turning to jelly and necks being wrenched as they jerked around to watch. Sometimes in the hall, she'd sneak up behind me and take my arm and let me walk her to class. It always made me feel proud that she picked me.

Roy Lee turned almost completely around in his chair. "Oh, Valentine," he crooned. "My *sweet* Valentine."

"Shut up, Roy Lee," she grumped and then turned her radiant smile back on me. "How ya been, honey?"

I never quite knew what to say to her. "I'm okay, thanks," I replied mundanely. "How about you?"

"Feeling fine," she replied while giving Roy Lee the sly eye.

"Only thing'd make it better if you and me went smooching some-where."

I melted into my chair while Roy Lee gleefully poked my ribs. "Forget Dorothy," he whispered. "Go get yourself some of *that*!"

I didn't get a chance to respond. A hush came over the audito-rium as Mr. Turner took the stage. He stood behind a lectern, his eyes darting angrily toward the slightest noise. Very quickly, even the most squirmy of us sat as quiet as stones. Then he spoke of two matters, each sufficient to shake the core of our young lives.

Big Creek High School, Mr. Turner said in his shrill voice, had been placed on football suspension for the 1958 season. That meant no games would be played. *None.* The reason for the suspen-sion was this: A group of well-meaning parents—the Football Fa-thers—had failed in their suit to force the West Virginia high-school athletic commission to let Big Creek play in the 1957 state-championship game. We all sat in dumb shock. He might as well have announced he was going to burn down the school. There was a groan from the knot of football boys. Coach Gainer stood up and hushed them. "Act like men," he said. "Show them what you're made of."

Mr. Turner had more to say. Big Creek was to be restructured, he said, beginning with the junior class. A more challenging aca-demic curriculum was to be installed, the result of *Sputnik* and the worry over how badly educated America's children were compared to Russian kids. Mr. Turner gripped the lectern and looked down on us. "There will be no more easy classes at this school," he announced.

We had heard two hard things, Mr. Turner said. "You can do nothing about the football suspension," he said, looking at the football boys. "Accept it and make the best of it. But the changes in the classroom are another matter." He gripped the lectern. "Af-ter you leave Big Creek, some of you boys will go to work in the coal mines, some will go into the service, some—not enough, in my opinion—will go to college. You girls will be wives, nurses, teachers, secretaries, maybe even someday one of you will be the president of the United States." There was a murmur of laughter, quickly smothered by dour looks from the other students.

Mr. Turner swept his gaze over us, his expression proud and certain. "The newspapers and television say the Russian students are the best in the world," he said. "They tell us how intelligent they are, how advanced, and how all the world may have to bow down to them when they take over. Well, I'm here to tell you Big Creek students have *nothing* to be ashamed of in front of *anybody*. You come armed with a wonderful education provided by caring teachers. You come from the best, hardest-working people in the world. You come from the toughest state in the Union. The Russians? I pity them. If they knew you like I know you, they'd be shaking in their boots!"

Our six hundred faces gazed up at the little man in rapt attention. Until a football boy stifled a groan, the silence was utter. Mr. Turner looked sharply in the football boy's direction, and Coach Gainer stood up and looked at him too. The football boys leaned their heads in on the groaner as if in group prayer.

Mr. Turner's eyes left them, went back to the general assembly. "Now, about these new standards," he said. "They're not going to be easy to meet. It's not only content. From my analysis of the new curriculum, there's at least twice the amount of material to cover in a school year. That's going to mean a lot of concentrated classroom work and a lot of homework.

"You must completely dedicate yourselves to it. To do less will be to let down your country, your state, your parents, your teachers, and, ultimately, yourselves. Remember this: The only good citizen is the well-educated citizen.

"Consider this poem by William Ernest Henley," he said, opening a book and adjusting his glasses.

"Ughhhh," Roy Lee growled, getting restless. I could sense a stirring in the student body. Mr. Turner had kept our attention until then. But a poem?

Mr. Turner's poem turned out to be *Invictus*. As he read, all of us, even Roy Lee, were absorbed by it. Mr. Turner concluded: " 'It matters not how strait the gate, how charged with punishments the scroll, I am the master of my fate: I am the captain of my soul.' "

He clapped the book shut. I almost jumped out of my seat. In the hushed silence, it had been as loud as a rifle shot. "Now the cheerleaders will lead us in the school song," Mr. Turner ordered.

The cheerleaders had been sitting together. They crept uncertainly on stage. They were not in their uniforms. "Sing," Mr. Turner ordered them. "Everybody *sing!*"

"On, on, green and white," the cheerleaders sang weakly, one looking at the other. The audience picked up the words, helping out. Soon, the whole auditorium was roaring. *"We are right for the fight tonight! Hold that ball and hit that line, every Big Creek star will shine! We'll fight, fight, fight for the green and white. . . ."*

When we were finished, there was applause and hoots of enthusiasm, almost as if we were cheering the team making a goal-line stand. But, with nothing to really cheer about, the noise quickly died and then there was a confused silence. Mr. Turner stepped from behind the lectern and nodded to our teachers, who stood up and began to shoo us out of the auditorium.

"Well, now, ain't that some shit," I heard Valentine say behind me as she made for the aisle. Roy Lee, like me, was too taken aback to say anything at all. The football boys gathered around their exalted coach, begging in vain for some reprieve. I looked for Dorothy and saw her with Emily Sue. Dorothy's cheeks were wet with tears. I wanted to go to her, but there were too many other kids in the way. By the time I made my way through them out of the auditorium, she had disappeared. As I exchanged books from my locker, Buck came up and smashed his fist against his locker's metal door. *"Dammit!"* he bellowed, and students stopped in their tracks, startled by his rage. Mr. Turner appeared instantly. Everyone but me and Buck fled the scene. I couldn't. He was standing in front of my open locker.

"Mr. Trant, I hope you didn't dent your locker," Mr. Turner said, his voice as cold as ice. "If you did, you will pay for its repair. And did I hear you curse? That I will not tolerate in my school, young man."

Buck, huge and thick-browed, loomed over the little principal. "I'll never get a football scholarship now," he said, and his lower lip began to tremble and a big tear welled up and dribbled down his fuzzy cheek. "It's the coal mines for me for the rest of my life. It just ain't fair!"

"*Isn't* fair, and you're correct. There was no fairness involved with this decision, only a petty sort of revenge. Even so, there will be no such displays in my school."

Buck frowned, his little eyes deeply puzzled. "But what will I do, Mr. Turner?"

"Do? You will do as we all do each day—our best with what God has given us. Now, if you are quite through with your sniveling, go to class." He leveled his hard little dark eyes at me. "And what are you looking at, *Mr. bomb builder?*"

Not a thing. I reached around Buck, grabbed my books, and took off at a near trot. "No running in the hall!" Mr. Turner called after me just as I turned a corner.

Afternoon classes were subdued. Dorothy dabbed at her eyes all through biology. When the bell rang, she gathered up her books and made for the door. I followed, but she was met just down the hall by Vernon Holbrook, a senior linebacker. Racked with sobs, she fell into his shoulder. He held her and then touched her cheek, wiping her tears away. Emily Sue walked up beside me and took it all in. "My, oh, my," she sang.

I could hardly breathe. "Don't say anything," I managed to growl.

"Wouldn't think of it," she said. "It speaks volumes all by itself."

"Listen, Emily Sue . . ." I was about to give her both barrels, but she walked away, heading down the hall toward our next class. When I looked back, Dorothy and Vernon were gone too. In a sea of jostling students, I felt all alone.

As soon as Jim hit the front door after school, gloom and anger descended on our house. He showed his displeasure by throwing his books on the living-room floor and stomping up and down the steps and slamming doors and yelling at Dad the moment he got home from work for causing the debacle. "That's enough, Jimmie," Mom admonished him while Dad stood in stricken silence.

"You've ruined everything!" Jim whined. "I won't get a college scholarship now!"

"You'll go to college," Dad said evenly. "I'll pay your way through. Don't worry about that."

"But I wanted to play college football! If I sit out my senior year, no college is going to pay any attention to me! I'll never forgive you, Dad!"

"James Venable Hickam, I said that's *enough*," Mom said, her tone of voice turning flat and hard. That was her warning tone. Jim opened his mouth and then clapped it shut, recognizing that Mom was on the edge. He stomped up the steps, making Daisy Mae jump out of his way, and slammed his bedroom door shut. Later, while Dad retreated to the living room and Mom fumed in the kitchen, the football boys from Coalwood gathered in Jim's room, plotting futile anarchy.

I had no sympathy for them. I was not even above stirring their pot just a little. I cracked open Jim's door and suggested, with no little glee, there might be room for them in the band. Jim sprang after me, and I ran back to my room and locked the door behind me. "You're dead, Sonny," I heard him say from the hall. A cold chill went up my spine, so solemnly was the statement made and confirmed by the cluster of huge, muscled boys standing outside my door. It was as if somehow every bad thing that had happened to them was now my fault.

A GLOOM seemed to settle over Coalwood. The fence-line gossipers mostly agreed that my father had acted stupidly. A common thread to the talk was that he'd gotten too big for his britches— again.

Dad didn't deliver any lumber and tin to Cape Coalwood, even though he had promised. After giving him a week, I decided to take direct action and went up to the mine carpentry shop to see Mr. McDuff. I entered the immaculate little shop, redolent of freshly sawn pine and oak, and found him working at a shrieking band saw. He shut it down, and I told him what Dad said I could have. He pushed his hand up under his white cloth cap and scratched his head. "News to me, Sonny. But there's a pile of scrap lumber behind the shop I guess is okay for you to take. You'll have to go see Ferro for tin. How's your mom like her new fence?"

As far as I knew, she liked her resurrected rose-garden fence just fine. It sure wasn't going anywhere. Mr. McDuff had built it back with posts as thick as telephone poles and crossbeams that could have been used as headers in the mine. The scrap lumber he sent me to look at behind his shop proved to be a stack of beautiful tongue-and-groove pine boards. When I asked, Mr. McDuff also slipped me a big box of nails. I called O'Dell, and a couple of hours later I heard the familiar rumble of the garbage truck pull up to the shop. We loaded the boards and then headed for the big hangarlike machine shops run in neomilitary fashion by Mr. Leon Ferro.

Rows of lathes, mills, shapers, and drill presses whined, ground, and hissed at us as O'Dell and I stepped inside. Twenty men worked during the day shift producing replacement parts for mine machinery and fabricating a variety of ductwork and support structures. When I asked for Mr. Ferro, the men at the machines waved me back to his office, a windowed cage overlooking the shop. Mr. Ferro leaned back in his chair, his hands clasped behind his head, and listened while I made my request for tin. "Sonny, I had some until this morning," he replied amiably, "but Junior Cassell came by and got part of it for a doghouse, and Reverend Richard got the rest to patch the roof of his church." He leaned forward. "Even if I had some, I don't give anything away out of this shop, even scrap. You want something, I like to trade. What you got?"

"Nothing," I admitted.

He shrugged. "Well, there you go. Come back when you do."

I decided to go and see Reverend Richard. We found him behind his church, pondering his little stack of tin. I wondered what he had traded Mr. Ferro for it. The Reverend was dressed in a black suit and a black tie, as if he had just come from a funeral. His shoes were black and white, long and narrow. He was holding a straw Panama hat. "Hi, boys," he said absently and then saw it was me. "Sonny boy! I sure miss the newspapers you gave me."

"I've missed your stories, sir."

O'Dell told him what we needed. "Love to help ya, I really would," he said, "but I don't have enough for my roof as it is."

I looked up. "But your roof is shingled."

He nodded. "If I had shingles, I'd use 'em. But I don't. I've got tin."

"Emmett Jones has a bunch of shingles stacked up next to his coal box," O'Dell said. "Almost the same color."

"Do tell," Little Richard said, suddenly interested. "I reckon I'd be up for a swap if you could manage it."

We were starting to figure out how to trade, Coalwood-style. We found Mrs. Jones pushing a lawn mower. "Emmett's at work," she said, eyeing the garbage truck, "but if you'll bring me a load of good plantin' dirt, those old shingles are yours."

The best "plantin' " dirt was down at Big Branch. We went by O'Dell's house, picked up two shovels and a pick, and kept going down past Cape Coalwood to where the road turned back up into the mountains. Beside a mountain stream, O'Dell and I worked at a clear place, picking and shoveling rich, black West Virginia loam into the truck. We were covered with dirt and sweat by the time we finished. Mrs. Jones was thrilled when we arrived with our load of dirt. "Oh, my flowers are going to be *glorious*!" she said as if she could see them already.

Just as the sun dropped below the western ridges, O'Dell dropped off our tin at the Cape beside the lumber and the nails. The next morning Quentin hitched over the mountains in time for breakfast. Mom made him eat an extra stack of pancakes. When he was done, he was so stuffed he could hardly walk. I raided the basement for hammers and saws and threw them in the back of Roy Lee's wreck of a car. We picked up Sherman and O'Dell on the way.

O'Dell had drawn up a plan for our blockhouse on a scrap of notebook paper. "I'm not the carpenter or the carpenter's son," he chanted as we sawed and drove nails, "but I'll do the carpentryin' until the carpenter comes."

The sun beat down on us, the slack yard a caldron of focused heat. To keep our morale up, we sang with discordant enthusiasm.

We went through the parts we could remember of "Be-Bop-A-Lula," "The Great Pretender," "Blueberry Hill," and "That'll Be the Day." If we didn't know all the words, we just repeated the ones we knew over and over again. Roy Lee had a good voice. With a sly eye in my direction, he gave us a solo rendition of the Everly Brothers's "All I Have to Do Is Dream," with modified lyrics:

"Dream, dream, dream,
all Sonny does is dream, dream,
   dream.
When he wants Dorothy in his arms,
when he wants her and all her charms,
whenever he wants Dorothy,
all he does is dream . . .
Only trouble is, gee whiz,
he's dreaming his life away . . ."

I laughed off Roy Lee's song, but it stung just the same.

When we got too hot, we went to the muddy little creek that ran behind the slack dump and sat down on the rocks and let the cool water run over our feet. Quentin, so hot he was feeling dizzy, stretched out in the creek, and we left him and went back to work. "We've got to have a launchpad too," O'Dell told us.

"Anybody here ever pour concrete?" I asked the group.

"I'm not the concrete pourer or the concrete pourer's son . . ." came back the cheerful chorus of replies as Quentin, staggering a little, wandered up from the creek to join us, complaining that a crawl-dad had bitten him. I declared the day's work at an end. We were all pretty exhausted.

When Mom saw Quentin, she admonished me for being heartless and cruel to the poor boy. She made him drink enough water to sink the battleship *Missouri,* fed him corn bread and beans, and then sent him to sleep in *my* bed. I got some spare blankets out of the hall closet and bedded down on the living-room couch for the night. Dad came in late and found me there. He switched on a lamp. "Heard you've been raiding my shops," he said.

I peeked above the blankets. "You said I could have scrap."

"I guess I did," he acknowledged absently, and then seemed to take note of where I was. "Why are you sleeping on the couch?"

"We spent all day building our blockhouse, and Mom said Quentin was too tired to go home. He's sleeping in my bed."

"You've already got your blockhouse built?"

"About halfway. You want to come see it?"

He yawned. "Just remember your promise. No sight or sound of any rockets in Coalwood."

"Yessir," I replied glumly. If Jim was a member of the BCMA, I thought, Dad would have been down there nailing boards with the rest of us.

"I saw Ike at the face today," he said, almost as an afterthought. "He said something about teaching you to work around a machine shop. I said okay, as long as it was on his own time and no company materials."

Mr. Bykovski had remembered! I grinned at Dad. "Thank you, sir!"

My enthusiasm took him off guard. "Don't get out of control now," he said.

"No, sir, I won't."

"No company supplies," he reiterated. "You understand? You can use the machines, but you've got to buy your own aluminum and steel."

"I've still got money from my newspaper route," I said. I was still grinning.

Dad looked at me with some confusion, almost as if he had never really looked at me before. "Good night, little man," he said finally, switching off the lamp.

"Good night, sir," I replied happily.

I cuddled back under my blankets and listened as he tiptoed up the steps. The living room was underneath my mother's bedroom. I heard the boards creaking beneath her feet as she crossed the room to her door as Dad crept down the hall. There were a few moments of silence, and then I heard her walk back to her bed and the sound of her mattress taking her weight. I heard Dad's bedroom door close. I guess I was starting to grow up, because for the first time I understood at least a little about the loneliness and frustration that often seemed to fill our house.

9

JAKE MOSBY

Auks V–VIII

EVERY YEAR, THE Ohio mill chose some of its young engineers and sent them down to Coalwood to a kind of coal-mining boot camp my father ran. The first thing Dad did with a junior engineer, as they were called, was to take him inside the mine and tramp him around for miles. The average height of the mine roof was five feet. To walk under it required a bent-over, head-up, forward-lunge kind of posture. The miners could always tell when Dad had one of his youngsters in tow, because they could hear them coming, Dad giving his running commentary on how the mine worked and the junior engineer's helmet bap-bapping off the roof. After a couple of days of Dad's torture, more than a few of them packed their bags and headed back to Ohio. One of them who stuck it out was Jake Mosby. Jake was to become important to the BCMA.

I first met Jake when I was in the ninth grade. Some of my newspaper customers lived at the Club House, the neo-Georgian mansion that sat on a small mount across from the Big Store. The Club House had been built for Mr. Carter's son after he returned from World War I. A housing shortage during a mine expansion in the 1920's had caused it to be converted into a boardinghouse. Since then, it had been gradually expanded until it had dozens of rooms for single miners or transient families.

Mrs. Davenport, the Club House manager, told me to go on up to Mr. Mosby's room. He had been there for a week, she said, so she guessed he was going to be around long enough to take the paper. I found Jake sprawled on his face in front of the door to his room. He wore the typical junior engineer's uniform: a canvas shirt with baggy khaki pants tucked into brown leather miner's boots. A foot away from one of his outstretched hands was an empty fruit jar. One sniff and I knew it had contained some of John Eye Blevins's moonshine. John Eye had lost a foot in the mine, and the company looked the other way when he supplemented his tiny pension dealing out fruit jars filled with the clear, fiery liquid. I placed an extra newspaper by the jar and started to leave, but Jake stirred. "Who are ya?" he demanded, his eyes still closed.

"Newspaper boy, sir," I answered. "Would you like to subscribe?"

Jake rolled over and sat up and then wiped his mouth with the back of his hand. He pushed the newspaper away and went for the jar, then threw it aside when he saw it was empty. "Dammitto-Christhell." He blinked at me and ran a hand through his sandy hair. "What time is it?"

"About ten after six, sir."

"A.M. or P.M.?"

John Eye's stuff had really done its work on this one, I thought. "It's morning."

He cursed again and tried to rise, managing only to get to his knees before dropping like a sack of potatoes. He curled up and held his stomach. "I'm dyin'," he announced with a groan that trailed off into a deep sigh.

"You want me to get Doc?" I asked.

He held up his hand and limply beckoned me closer. "No doctor. What's your name, boy? I'd like to know who I'm with when I go to the angels."

I told him and then shook his damp hand when he stretched it out to me. When he let go, I wiped my hand on the back of my jeans.

"No kin to Homer Hickam, are you?"

I told him.

"Your dad . . ." he began, "your dad . . ." He searched his scorched brain for just the right words and rolled over on his back and flopped an arm across his eyes. "Your dad . . ."

"I've heard my dad's a sonuvabitch when it comes to you Ohio junior engineers," I finished for him, as dryly as my age and the hour allowed.

Mosby laughed. "Oh, oh, that hurts." He raised his head, only one eye open. "You're right, Sonny. He is one mean SOB."

"Welcome to Coalwood," I said. "Want to take the *Telegraph*?"

He didn't, said he couldn't afford it, and I took the complimentary copy with me. I mentioned him to Mom afterward. She laughed. "You want to know something else about your Jake Mosby? His daddy owns about twenty percent of the steel mill that owns us. He's got more money than Carter's got little liver pills."

I next saw Jake at the company Christmas party at the Club House. It was the first time I'd actually seen him standing. He was leaning on the mantel of the fireplace in the big hall on the first floor, a drink in his hand, talking to Mr. Van Dyke's new secretary, a pretty, pert blonde imported all the way from New York City. Mosby was dressed splendidly, in a tuxedo, the first one I had ever seen. He was tall—loose-limbed, as my mother would say. She and some other ladies were in a corner, eyeing him. "He looks just like Henry Fonda," I heard one of them say. The secretary was also part of the ladies' conversation. "Have you heard that accent?" one of them tittered. "Nyah, nyah, nyah. How do they understand each other up there in the north?"

Later that night, Jake enticed the secretary outside for a ride in his Corvette. Even over the loud, discordant music being performed by Cecil Sutter and the Miners, I heard her delighted shrieks as he spun the car around on the ice-coated road that ran down beside the church. By the time the two of them made it back to the Club House, they were, as my mother sniffed, "drunk as Cooter Brown." Jake and the secretary proceeded to the dance floor and did a dirty dance, people falling back from the pair in shocked silence. The band wound down and the accordion player's mouth fell open when Jake wiggled down behind the secretary, coming close to smooching her behind right there in front of God and everybody. He stood up. "Why'd you stop the music?" he slurred. He leaned on the table loaded with desserts, but it collapsed and he fell with it, everything sliding down on top of him.

He lay there, his face covered with a stupid smile and red and green cake icing, until my dad ordered him dragged out by his legs. He was left semiconscious on the porch steps, the new snow covering him, until I convinced Jim to help me get him upstairs to his room. Mr. Van Dyke's imported secretary left town the day after New Year's. Jake stayed on because, as Dad explained to Mom, Mr. Van Dyke thought he "had promise."

"Well, of course, Homer," Mom replied, not entirely successful at stifling a laugh. "I'm sure it couldn't have anything to do with who Jake's daddy is, now, could it?"

Jake was a hiker, and since I knew every nook and cranny in the surrounding mountains, he occasionally called me up and paid me to guide him and whatever girlfriend he had at the time. Jake had been a fighter pilot in Korea and had been all over the Orient. "Oh, man, we blazed through the wild blue yonder over there," he told me one time up in the woods while his girl was behind a bush watering the daisies. "Almost got me a MiG. Missed the sonuvabitch by that much. I don't know how many whores back in town it took for me to get over that."

I was more impressed by the reference to women than his nearly bagging the MiG. "How many women have you been with, Jake?" I asked him.

He howled with laughter. "I'll tell you if you'll tell me."

I held my fingers to make an *0*.

"Well, gawdalmighty, Sonny," he said, shaking his head. "I thought the definition of a West Virginia virgin was one who could outrun her brother. What's your problem?"

"It's 'cause he's a little gentleman, Jake," his girl called out from behind the bush. "Unlike you, I might add."

"Wisdom from the outdoor toilet," Jake laughed, rolling his eyes.

I envied Jake and his ease with women and wondered if I would ever learn to be the same. I sincerely had my doubts, considering how tongue-tied I got around them sometimes. "Don't worry about it, Sonny," Jake said when I expressed my lack of prospects as far as women were concerned. "There are two things every woman really wants: one, she wants to know that a man really loves her, and two, that he isn't going to stop. Unlike me, more's the pity, you got the makings to be that kind of man. When they figure that out, the girls are going to be after you."

For all our differences in age and outlook, Jake and I became friends. He unfailingly sought me out at the Big Store when he saw me to ask me how I was doing and, lately, what was going on with the rockets. When I told him about our progress, he flattered me by promising to come down to Cape Coalwood and have a look for himself. I hoped that he would.

THE summer of 1958 came, and with it, floating ships of clouds that lazily drifted by, docking sometimes in the afternoon to produce a shower to loosen the dust off the houses and the cars. Katydids sang their repetitive song in the evenings, and rabbits came down from the mountains to investigate the dozens of little tomato and lettuce farms along our steep hillsides, taking their chances with Daisy Mae and Lucifer. At night, as the stars unfurled, cool air cascaded off the hills into the valley. I often went out into the yard after dark and lay down on the grass and looked up into the sky, hoping to catch a glimpse of a satellite going over. I didn't see any, but I had fun looking all the same.

In May, the company announced that its big new coal-preparation plant in Caretta was complete, and all the coal from both the Coalwood and Caretta mines would henceforth be loaded into coal cars over there. It took awhile before everybody realized what exactly that meant. Coalwood was in for a major change. The Coalwood tipple would no longer lift coal out of the mine, and no more trains would go chugging through town or spewing dust off the coal cars. Dad said to Mom at supper one night that even the tracks were to be taken out. This announcement was not greeted with overwhelming joy. Some Coalwoodians saw a conspiracy in the whole thing. Roy Lee said the union was afraid this was the start of the end of the Coalwood mine. If everything could be done at Caretta, who needed Coalwood?

At Cape Coalwood, we needed concrete for our launchpad. O'Dell scoured the town and came up empty. That meant I had no choice but to go ask Dad for his help.

Mr. Dabb, his clerk, said he was inside the mine, so I waited at the shaft as miners went up and down on the lifts. There were two lifts, or cages as they were sometimes called, side by side. When one went up, the other went down. The one that was up was kept about six feet above ground level. That was so nobody could get on without the hoist operator knowing about it. Miners wanting to go down the shaft pushed a brass button beside the cage, which rang a bell. One ring told the operator to lower the cage into place. Two rings meant the miners were getting aboard. Three meant to "Bail 'em out," or lower the cage.

Mr. Todd took care of the lamp house, where the batteries for the helmet lamps were charged. It was also his job to inspect each man before he got in the cage and make sure he didn't have any matches (the Coalwood mine was notoriously gassy), had his helmet on, and was wearing hard-toe boots. While drinking a bottle of pop Mr. Todd brought me, I watched the miners go through their routine of coming and going. Each miner had two brass medals with a number stamped on it. To get a lamp, a miner presented one of his medals, which Mr. Todd hung on a board. The other medal went in the miner's pocket. A glance at the board told my dad or anyone else who was in the mine. The medal the miners carried with them provided identification in case they were hurt or killed. It was no secret in Coalwood that injury and death was always a possibility for every man every day in the mine, no matter how hard Dad and his foremen worked to keep things safe.

When I was in grade school, every so often one of my friends would be called from class and not return, and I would learn at supper that my friend's father had been killed in the mine. This fact was normally presented matter-of-factly by my mother. My father rarely told us any details. I'd get those later, from my friends at school. Once, when I was in the fourth grade, a little girl with golden curls named Dreema was called out of class. I never saw her again. Her dad had been beheaded by a sharp piece of slate when the tunnel he was working in collapsed. Dad came home from work that night with his hands bandaged, bloodied from removing the rock that trapped the men. He fired the foreman responsible for failing to properly support the roof of his section. After that, no one said any more about the incident. The company required the dead miner's family to move within two weeks of the accident that killed him. Perhaps deliberately, there were almost no widows in Coalwood to remind the rest of us what could happen in the mine.

Mr. Dubonnet and a knot of other miners were gathered in an impromptu union meeting by the lamp house. He was handing out pamphlets. "I heard you got your rockets flying," he said to me.

"How high are they going?" one of the other miners wanted to know. "You hit the moon yet?"

"Come and see," I told him.

"When are you shooting them off, Sonny?" Mr. Dubonnet asked. "I'd like to come down and watch. Bet a lot of people would."

I had a sudden inspiration. "I could put up a notice at the Big Store and the post office."

The lift bell rang twice, and the men shuffled aboard. "I'll be there," Mr. Dubonnet said as he descended.

Dad came up on the return lift. Before he saw me, I watched him as he pulled out a red bandanna, now gray, and coughed into it and then spat into a pile of gob next to the bathhouse. He looked up and waved me in behind him as he went into the bathhouse. He hung his helmet on a peg, stripped off his boots and coveralls, and got into the shower and started lathering up with Lava soap. "Why are you here?" he demanded while attacking the black grime embedded in his face.

"Could I please have some cement?"

"No," he answered. A puddle of coal mud swirled about his feet. "What do you want it for?"

"We need a launchpad. I thought if you had some cement that was maybe extra—"

"The company doesn't have extra cement," Dad muttered

through the spray, twisting a washcloth into his ear. "The company doesn't have extra anything. If we did, we'd go out of business. How many bags do you need?"

"Maybe four?"

Dad finished and toweled down. I knew he'd take another shower when he got home to scrub more coal off. The coal dust that collected in the moist skin around his eyes would remain—the miners of Coalwood walked around with their eyes lined like Cleopatra's. "I tell you what," he said as he toweled off. "I had a junior engineer make the estimate on a walkway up at fan number three, and I heard there was some cement leftover. It's rained since then so it's probably ruined, but you can have it if you want it. Save the company the expense of hauling it out."

He didn't have to tell me twice. The next day, after his garbage run, O'Dell borrowed his dad's truck and he and Sherman and I went up the tortuous trail to one of the big fans that drove air through the mines. There, beside the locked door to the fan controls, sat four bags of cement. They hadn't been rained on at all. There was also a pile of sand and gravel, equally intact.

"Are you sure your dad said we could have this?" Sherman worried. "It's prime."

I shrugged. "He said the rain ruined it."

"What rain?" O'Dell demanded. "It hasn't rained in a month. Your dad's fooling with you, Sonny. Look, there's the new walkway. All done. They could've hauled the cement and stuff away when they finished."

I considered the implication of what O'Dell was saying. Was Dad helping us? Or maybe he'd made a mistake because he was so distracted by the football suspension and opening the new preparation plant over in Caretta. God only knew, but I didn't have time to figure it out. "Come on," I said, "let's load it up before somebody beats us to it."

AFTER we dug a hole in the slack and poured a five-by-five-foot slab of concrete for our launchpad, Cape Coalwood was ready for its first rocket. The blockhouse was thirty yards away from the pad, on the creek bank, its dimensions determined by the lumber at hand. Quentin grandly described it as an "irregular polyhedron," but it was little more than a wooden shed. It had an earthen floor, a doorless entrance in the back, a flat tin roof, and, for a viewing window, a wide rectangular opening covered by a clear, quarter-inch-thick sheet of plastic that O'Dell found, slightly scratched, in the trash behind the Big Store. Mr. Dantzler used these sheets to protect his glass counters. Beside the blockhouse, we erected a flagpole, of two-inch galvanized pipe discovered abandoned alongside a gas wellhead up Mudhole Hollow (Mr. Duncan, the company plumber, told me about it). A BCMA flag, sewn and stitched by O'Dell's mother, fluttered proudly from it. I loved the flag. It had the initials *B-C-M-A* arched over an embroidered rocket with an owl (the high-school mascot) riding on it.

To open Cape Coalwood, I loaded *Auk V* with our bottle-tested formula of finely ground black powder and postage stamp glue, cured under the water heater for five days. Because I had promised Mr. Dubonnet and the other miners at the tipple I'd let them know when there was going to be a launch, Sherman posted a notice on notebook paper in big block letters on the bulletin boards at the Big Store and post office:

ROCKET LAUNCH!

THE BIG CREEK MISSILE AGENCY (BCMA)

WILL LAUNCH A ROCKET THIS SATURDAY, 10:00 A.M.,

AT CAPE COALWOOD

(THE SLACK DUMP TWO MILES SOUTH OF

FROG LEVEL)

True to his word, Mr. Dubonnet came to our next launch, parking his Pontiac at a wide spot on the road opposite our blockhouse. There was usually a union meeting on Saturday morning, so I knew he had to hustle to make it to the Cape on time.

I was pleased when Jake Mosby also showed up, driving his Corvette. Tom Musick, another junior engineer, was with him. After carefully parking his car under a protective tree, Jake sat down beside Mr. Dubonnet on the Pontiac's fender and raised a bottle of beer in my direction. Tom just waved.

I was surprised to see another car drive up. It was an Edsel driven by a man named Basil Oglethorpe. Jake, as it turned out, had invited him. He waved me over to introduce us. Basil had the physique of Ichabod Crane. He had on a cream-colored suit, a wide-brimmed floppy hat, a black string tie, a silk vest, and narrow shoes that had weaving in the toes. He also wore a fob watch and a chain. I had never seen anyone so quirkily dressed in my life. My mouth dropped open at the sight of him. Basil ignored my reaction, one that he was probably used to getting in McDowell County, and told me he was going to make me and the other rocket boys famous. "I'm going to be your Lowell Thomas, Sonny my boy," he told me, "and you my Lawrence of Araby."

"Basil's with the *McDowell County Banner,*" Jake said, watching my reaction. He was clearly amused. "It's a grocery-store rag."

"We're growing, however," Basil sniffed, taking a big flowery silk handkerchief from his vest and pressing it to his nose. "I am the editor-in-chief and features writer."

"He sweeps out too," Jake added. "I thought he might help you boys get some attention. Seems to me you deserve some, as hard as you've worked down here at this old dump."

I wondered how interesting we would be to a real writer. I couldn't imagine it. I shrugged and went back to supervise preparations for the launch. Roy Lee lit the fuse to our little *Auk* and ran for the blockhouse. Before he got there, the fuse reached the powder and the rocket whooshed off the pad, climbed about fifty feet, and then, as if aimed, turned and flew directly at the men lounging on the Pontiac fender. Mr. Dubonnet, Jake, Tom, and Basil threw themselves to the ground while the rocket hissed overhead and then slammed into the road behind, skittering along until it plowed into a muddy ditch. It happened so quickly I didn't have time to react. "Damn! Never saw men move so fast in my life," Roy Lee observed.

We chased after the rocket. Sherman stopped long enough to help Mr. Dubonnet and Tom up. Basil was whooping and laughing and dancing around, stopping to scribble on a notebook pad. "Oh, it's just like Cape Canaveral," he exclaimed. "I love it!"

Jake had gotten up on his own and walked rapidly down the road. I watched as he lit a cigarette with trembling hands and took a pull off a flask. I went to see if he was all right. He waved his cigarette around. "Seeing that rocket come at me was almost like being back in Korea," he said shakily.

"I'm real sorry, Jake" was all I could say.

"It don't mean a thing," Jake said. His fingers brought the trembling flask to his mouth.

When I came back up the road, Mr. Dubonnet and Tom were inspecting the rocket with the other boys clustered around him. Basil was in his Edsel, still scribbling furiously. "Boys, next time I come down here, I'm going to make sure my insurance is paid up." Mr. Dubonnet ho-hoed. He sniffed at the rocket. "Your powder's putting out a lot of tailings. This is black powder, right?"

Our own special mixture, I told him. Mr. Dubonnet tapped the rocket casing, and lumps of unburned propellant and ash dropped out. He smeared some in the palm of his hand. "Still wet," he said. "How long did you let it cure?"

I told him five days.

"I'd give it at least two weeks, Sonny." He rubbed more of the powder residue between his fingertips. "I worked explosives before the company brought in the continuous-mining machines. Powder's got to be bone-dry."

After Mr. Dubonnet, Tom, and Jake left, we boys gathered with Basil beside the blockhouse to discuss the results of the flight. "We've got to figure out how to make our rockets fly straight," Sherman said.

"And we've got to find a better way to set them off," Roy Lee observed, well aware of what might have happened if the rocket had veered after him on the path back to the blockhouse.

Quentin said, "I'll think about it, come up with some proposals."

"I hope you come up with something better than that stupid test stand," O'Dell said.

"O'Dell, cut it out," I broke in. "We're a team here, remember?

Quentin, work on it. We'll get back together after miners' vacation. All agreed?"

"Hell, yes," Roy Lee said. "Did you see the way our rocket flew? So what if it didn't go straight? We're doing good here!"

"Roy Lee's got it right," I said. "We're making progress." I put out my hand, palm down. "Come on. Put your hand on mine, like the football team does."

One by one, Sherman, O'Dell, Roy Lee, and Quentin solemnly placed their hands one on top of the other, all on top of mine. "Rocket boys," I said. "Rocket boys forever!"

"Oh, this is so *perfect*!" Basil chirped and kept writing. "Rocket boys forever. *I love it*!"

THE Coalwood mine, just like all of the mines across McDowell County and the southern part of the state, shut down for the first two weeks of July, so everyone was required to take their vacations at the same time. My father said it was done that way so the economic clout of the mining industry would be clear when all the miners showed up on vacation at the same time. It was for the same reason, he said, that miners were often paid entirely with two-dollar bills, so that local merchants would realize how important the coal companies were to their businesses. Whatever the reason, Coalwood became almost deserted during miners' vacation. Hungry Mother State Park in nearby Virginia was a popular destination for miners and their families, and also the Smoky Mountains farther south in Tennessee. Another traditional miners' vacation spot was Myrtle Beach, South Carolina. At Mom's insistence, that's where we went. It was the one time of the year she got Dad off to herself and out of the mountains. Days would pass at the beach without Dad saying anything about the coal mine. I noticed that Mom often reached out and just touched his hand when he was talking, and when they sat in the swing on the motel porch at night, sometimes he would even put his arm around her shoulders. They even slept in the same bed. Once when I came back from trying to catch crabs with a fish head on a string, our motel door was locked. I knew my parents were in there because their sandy shoes were on the porch, but no amount of knocking got them to let me in. I guessed they were taking a nap. Mom cried when we loaded up the Buick to come home.

As soon as Dad drove into the backyard, we could hear the black phone ringing. "Welcome to Coalwood," Mom muttered to his back as he ran up the porch steps to answer it.

I'd left three loaded rockets, *Auks VI, VII,* and *VIII,* to cure in the basement while we were on vacation and decided to fire all three the Saturday after we came back. Sherman made up some handbills to post at the Big Store and the post office. Since we still had a guidance problem, I spent the next couple of days in the basement, tinkering with fins and how to attach them. O'Dell had delivered a thin aluminum sheet he'd found in the garbage, so I used some tin shears to cut some rough triangles out of it for fins. I punched holes along their inner edge with a nail and then used steel wire to strap the fins to the sides of the casements. After I cranked the wire down with pliers, the fins, though crude, appeared to be at least attached securely. I hoped they might do. On Saturday, Roy Lee came by in his car and I put the rockets in the front seat with us for the ride down to the Cape. He admired the new fin apparatus and said, "I wonder if we're going to have another crowd today."

Before I could stop myself, I said, "I wish Dorothy would come over."

He shrugged. "Why don't you invite her?"

"I'm afraid she'd bring a boyfriend," I answered honestly.

"She's going out with other guys and you're still mooning over her?" Roy Lee shook his head. "Sonny, you and I have got to sit down and *talk.*"

"I love her," I said, "and someday she's going to love me."

He sagged behind the wheel, shaking his head. "That ain't the way it works, boy."

At the Big Store, men sitting on the steps waved at us. "Rocket boys!" they hallooed. The football boys had taken to walking around in a gang all summer, as if daring anyone to say anything about their suspension. Buck, brother Jim, and the rest of the giant boys gathered in front of the Club House. They glowered at us as we passed them, but said nothing. We picked up Sherman and kept going. O'Dell was waiting for us, having walked down to the Cape from Frog Level. He had already cleaned out a hornet's nest in the blockhouse and swept off the launchpad. Mr. Dubonnet didn't show this time, but Jake and Basil were there. "I've been preparing my story on you," Basil said. "Just you wait and see the power of the press."

We let Jake light the fuse. He ran laughing to the blockhouse, his long legs pumping. This time, I had lengthened the fuse so he had time to clamber inside with Basil and the rest of us, all hunkered down in happy anticipation. Mr. Dubonnet had been correct about letting the powder cure longer: The rocket leapt off the pad with a louder-than-ever hiss and then streaked nearly out of sight. I climbed out of the blockhouse and caught sight of the wisp of its contrail as it fell downrange. I joined the others in a stampede of joy. It was our best rocket yet. "How high did it go?" Jake asked breathlessly, as excited as if he were a rocket boy himself.

"Twice as high as the mountains," Sherman said authoritatively.

How high was that? We didn't know.

"Maybe a little trigonometry would help you figure it out," Jake said.

We didn't know anything about trigonometry. "I'm a little rusty," Jake said, scratching his head. "But let me think on it."

Auks VII and *VIII* didn't need any trigonometry. *Auk VII* did a horseshoe turn not more than fifty feet up and slammed into the ground. *Auk VIII* bounced once in front of the blockhouse and then exploded overhead, rattling the tin roof with steel shrapnel. "Oh, this is so exciting," Basil exclaimed.

"Every time I get around you boys, it's like being back in Korea," Jake said, marveling at the gouges in the blockhouse. "The Army's going to love you—if you live long enough to join up."

The next weekend, Mr. Bykovski met with me and Sherman and O'Dell at the tipple machine shop to teach us the fundamentals of welding and cutting steel so that we could build our own rockets. I wondered if Dad had agreed to it because of his guilt for banishing Mr. Bykovski to the mine. I couldn't imagine it to be true, but I wondered it all the same. Mr. Bykovski claimed to be happy in his new job. I counted out five U.S. dollars to pay for the steel tubing and bar stock we were to use that day. I left it, according to Mr. Bykovski's direction, on the workbench with a note on what it was for.

Dad stood on the stoop of his office when we came out after the training. I went over to thank him. "So I guess you're an expert at welding and machining now," he said.

"No, sir," I replied. "It'll take a lot of practice to be any good at it at all."

He looked a little surprised at my response and then nodded. "Not much comes easy in this world, Sonny. If it does, it's best to be suspicious of it. It's probably not worth much."

"What's the hardest thing you ever learned, Dad?" I asked abruptly.

He leaned on the rail of the stoop. "Entropy," he said finally.

I didn't understand the word and he knew it. "Entropy is the tendency of everything to move toward confusion and disorder as time passes," he explained. "It's part of the first law of thermodynamics."

I must have looked blank. "No matter how perfect the thing," he continued patiently, "the moment it's created it begins to be destroyed."

"Why was that so hard to learn?"

He smiled. "Because even though I know it to be true, I don't want it to be true. I *hate* that it's true. I just can't imagine," he concluded, heading back inside his office, "what God was thinking."

ONE evening that summer, just before school started again, Jake called me on the black phone. Dad took the call, of course, and handed it over, a suspicious look on his face. "Make it fast."

"Sonny," Jake said, "you boys come down and join me on the roof of the Club House tonight. Got a surprise for you."

Sherman was the only rocket boy I could round up on short notice. I scampered up the ladder to the Club House roof. Sherman doggedly hopped rung to rung on one foot.

Jake looked up from the eyepiece of a long cylinder pointing skyward. "A beauty, ain't it?" he grinned proudly. "My old trusty refractor. Just came in the mail today. This is what I used to do when I was your age, boys. I'd almost forgotten about it until I came down to your range." It was the first telescope I had ever seen. He handed me a battered book. "Had my mother mail me this too. My old trig book. Learn this stuff and you can calculate how high your rockets fly."

The night was clear and the stars were spread out like diamonds on a vast blanket of black velvet. "Come on, be my guest." Jake grinned. "I got Jupiter cornered."

Sherman went first, pressing his eye against the eyepiece. "I can see the bands!" he cried.

I took off my glasses and Jake showed me how to rotate the focus knob. Jupiter was a shimmering yellow circle with brown horizontal streaks. It felt as if I could reach out and touch it, and I wanted to.

Jake pointed up at a stream of stars snaking across the sky between the mountains. "That's the Milky Way, our galaxy. We're looking at its edge." I heard him unscrew a bottle and take a drink. He whistled out a long breath. "That's the constellation Lyra the Lyre, and there's Sagittarius the Archer. But look there, beside Lyra." He fumbled with the eyepiece. "Tell me what you see."

Sherman looked and then I did. A glowing doughnut. I could just barely make it out. "A star with a hole in it?"

Jake laughed. "Close. It's the Ring Nebula. The ring is the ejected shell of a star's outer mass."

Long past midnight, Jake kept showing Sherman and me different planets and stars, until finally he sat down and leaned up against a brick chimney and went to sleep. While Sherman kept looking through the telescope, I wandered to the edge of the roof and looked out over my little town. The church, bathed in starlight, glowed against the black silhouette of the mountain behind it, and on the hill above the post office I could make out the spires of Mr. Van Dyke's mansion. The trees rustled in the cooling air coming off the mountains, and in the distance I could hear the hoot of a lone owl and by the creek that ran alongside the machine shops the rhythmic *eeping* of frogs. I went back to the telescope and tried to use it to look at Coalwood, but discovered I couldn't focus it close enough. I thought how ironic it was that Jake's telescope could see stars a million light-years away, but not the town it was in. Maybe I was that way myself. I had a clear vision of my future in space, but the life I led in Coalwood sometimes seemed to blur.

Sherman gasped so loud it made me look up in time to see the streak of a big blue meteor, yellow sparks flying from its head, coming out of the north. It flew silently across the sky and then fell behind a mountain. I wanted to say something to capture the glory of its passing, but I had no words that were adequate. Sherman and I looked at one another. *"Wow"* was all we could say. Jake kept snoring.

10

MISS RILEY

Auks IX–XI

I have seen the future and it works! Two weeks ago, this reporter watched as the boys of the Big Creek Missile Agency launched their magnificent creations at their new Cape Coalwood range. As their silvery missiles leapt from the concrete pad and soared away into the sky, my mouth dropped open, so enthralled was I at the glorious sight of their rockets scrambling toward space. . . . They also have their failures. I got to hunker in their bunker and dodge shrapnel with these brave lads. But they are not the kind of boys who give up! This reporter is telling one and all who read these words: If you have any hope of understanding what the grand and glorious future holds for all who dare seize it, you must come to see the rocket boys of Coalwood.

—*The* McDowell County Banner, *August 1958*

THE FIRST DAY back to school in 1958 also began the first day of the football suspension. Instead of swaggering heroically through the halls in their green and white letter jackets, Jim and the football boys trudged to class sullen and trigger-sensitive to insult. Usually, at the beginning of the academic year, the team would be nearing their first game and the school would be focused on them. All they had to do was crook their fingers and the girls would come running, eager to be known as the girlfriend of a member of the exalted Big Creek team. This year, it seemed they looked more lumpy than muscled, more bullet-headed than bright, and oddly tainted. They were still quite capable of wiping the floor up with me, so I kept my distance and recommended that the other rocket boys not tease them either. "But it's so tempting," Quentin snickered as we walked down the hall. "Look at them. Like lost sheep."

We were soon to learn that more had changed at Big Creek than the lack of football. Our teachers sat us down, shut us up, and began to talk rapidly at the blackboard, outlining the courses and what would be expected of us in the new *Sputnik*-inspired curriculum. Astonishing homework assignments filled our notebooks. Books began to stack up. Mimeographed handouts flew down the aisles. Clutching books and papers, we slogged from class to class, our arms wrapped around the material. The same thing was happening in high schools in every state. *Sputnik* was launched in the fall of 1957. In the fall of 1958, it felt to the high-school students of the United States as if the country was launching *us* in reply.

"Hi y'all," a pretty tenth-grade girl said to Quentin and me in the hall between classes. "Will you be at the Dugout this Saturday? Hope so. I love to dance." She skipped past a cluster of football boys without so much as a glance. They looked at Quentin and me with murder in their eyes.

"Wow," Quentin said. "That's never happened before."

"We've never been written up in a newspaper before either," I reminded him.

As we passed the trophy case, I saw Valentine. She was standing alone, her books held to her chest. She was wearing a plaid dress and a tight black sweater, and her hair was tied back in a ponytail, an ebony, shimmering waterfall. She looked sort of doleful. "Hi, Sonny," she said, her eyes lighting up at the sight of me. "You wanna go to the band room and neck?"

I was sure Valentine was only kidding. After all, she was in the class ahead of me and nearly two years older. I came over to her. "Sure, Valentine," I joked. "Any day, any time."

She seemed to search my eyes. "You want to walk a girl to class?"

"You bet."

Valentine leaned into me while we walked down the hall. "Read about you in the newspaper," she said. "I am so proud of you. Um, would it be okay if I and some of the other girls came over to see you launch your rockets?"

Valentine forever had the capacity to surprise me. "I'd be proud to have you," I told her, and it was the truth.

A knot of sullen football boys trudged past us, giving us dirty looks. One of them, Bobby Joe Shaw, bumped Valentine so hard she almost dropped her books. She grabbed his arm and spun him around. "Watch where you're going, snot for brains!"

Like Valentine, Bobby Joe was a senior. I remembered seeing the two of them holding hands in the auditorium the previous year. He had played second-string quarterback during the 1957 season, but he threw a mean pass and had a great running game too. His year to shine was supposed to be 1958. Now it was gone forever. "Robbin' the cradle, ain't you, Valentine?" he said.

"Don't give me any of your shit, Bobby Joe," Valentine snarled, nearly standing on the boy's toes. He backed off, glanced at me with a look that could kill, and then stalked down the hall. She came back. "Bobby Joe or any of those bad boys ever give you any . . . *junk,* Sonny, you come get me. I'll take care of 'em." At her classroom, she gave me a coy smile. "Whenever you're ready to make out, just call me. I'll be there." She gave me a wink and went inside.

Quentin came up beside me and helped me watch Valentine to her desk. "That is the most prodigious girl in this school!" he pronounced. Except for my Dorothy, I had to agree. My heart was thumping in my chest like I'd just run a mile. Coach Gainer had warned us boys in health class about the hormones that surged through our bodies when we got in high school. "It'll pass," the great man had advised. "Enjoy the sensations while you can, but don't act on them. If you realize they're not your brain talking, but just teen-age-boy crazy hormones, you'll be fine."

At the end of the day, Dorothy hailed me outside as I headed to the bus. She was a vision in a white starched blouse and navy blue skirt. "Coming over Sunday?" she asked. "I need help on plane geometry."

"I'll be there."

142 / ROCKET BOYS

She looked away, a little smile playing. "I missed you all summer," she said in a soft voice.

"R-really?" I stammered.

"Um hmm," she nodded, her big blues fastened on me. "I read about you too. All these cute little tenth-grade girls have got their caps set for you, I'll bet. I'm just as jealous as I can be!"

I grinned like a complete moron at her. "Don't be! I-I mean . . . Dorothy, I missed you too!"

"There's so much for us to talk about. I just can't wait!"

Roy Lee came after me at a trot. He stopped and eyed Dorothy with obvious distaste. He just never tried to understand her perfection. "Sonny, Jack's ready to go. He said you've got about five seconds or he's going to leave you here with Miss Priss."

Reluctantly, I followed Roy Lee. "I was hoping you'd be over her by now, Son," he said.

"Never," I replied.

I waved at Dorothy from the bus. She waved back and then threw me a little kiss. I felt like I was floating the whole trip back to Coalwood. Jack had to remind me to get off at my bus stop.

I had survived algebra in the tenth grade, barely managing a B after a flurry of good test scores at the end of the school year. But in the eleventh grade, I got good scores in plane geometry from the start. For one thing, I was certain its body of knowledge concerning plane curves, angles, and polygons would help me design my rockets. I suspected there were dimensional relationships involved in rocket design, such as a proper ratio between the area of the guidance-controlling fins to the area of the casement. But how could I figure out such things? Mr. Hartsfield waved away my questions about how to calculate and compare flat areas (the fins) and curved surfaces (the casement) to first immerse us in Euclidean geometry and all of its axioms and postulates and proofs. "Sir, you are asking questions that are more in the nature of analytical geometry and calculus," he said, turning away from the blackboard to eye me over his half-glasses. "As I recall, you had trouble understanding algebra. And if you didn't understand algebra, Mr. Hickam, you're lost, lost for all time!"

During a lecture on triangles, I had the sudden insight that there was a relationship between the three sides and the angles they formed. I asked about that and Mr. Hartsfield looked me over, but not entirely disapprovingly. "That, Mr. Hickam, is trigonometry. In due course, we will get to wherever it is your usually less supple mind is trying to take us."

My usually less supple mind was trying to figure out how high our rockets were flying. I delved into Jake's book. Quentin, delighted to have it, did the same. Sitting together in the Big Creek auditorium at lunch, we taught ourselves trigonometry. I had discovered that learning something, no matter how complex, wasn't hard when I had a reason to want to know it. With trig under our belt, all we would need to do was build some instruments to measure angles and we would be able to calculate how high our rockets flew. "I'll get right on it," Quentin promised.

"Oh, Sonny, you're so *smart*," Dorothy sighed on the couch in her living room when I told her how I was learning trigonometry. She leaned over and hugged me. "That's for helping poor me on this old plane geometry."

It was the perfect opportunity for a patented Roy Lee kind of move. I started to slide my hand around her shoulders, but she jumped up. "Oh, my cookies are going to burn! Be right back." When she returned with a plate of chocolate chip cookies, she sat in the chair across from me and doled them out. "I'm so glad we're friends," she said for about the millionth time. I didn't let it wear me down. I was making progress with her, one little step at a time.

All through the fall, I thumbed over to War every Sunday afternoon so Dorothy and I could work on plane geometry. We worked well together. As we covered each postulate and theorem, it soon became clear that Dorothy actually understood their derivations better than I did. She was a good teacher, patiently explaining to me how each proof built on the other. She had a wonderful memory for details and never seemed to forget anything once she had committed it to memory. But I was a lot better than she was at mental visualizations. I had to draw a picture for Dorothy just to get her to understand that two lines were parallel if they were both perpendicular to a third line.

Mr. Hartsfield did his best to give us a tool to do our work. "Ladies and gentlemen, you must learn deductive reasoning!" He caught Roy Lee ogling the girl next to him and threw a perfect chalk strike to the boy's head. "Now, sir, let me put a general statement to you," he said to Roy Lee. "All human beings have brains, that's my major premise. Do you not agree?"

Roy Lee rubbed his head, chalk dust sticking to his lacquered D.A. "Yes, sir."

Mr. Hartsfield stood and balanced on his toes. "And all teenage boys are human beings. That is my minor premise, controversial though it may be. And if my major and minor premises are so, sir, what is your conclusion?"

Roy Lee wrinkled his brow. "That all teenage boys have brains?" he finally allowed.

"Why, yes, my boy!" Mr. Hartsfield shouted and bounced a foot off the floor. "So what, pray tell, is your excuse?"

Deductive reasoning was all well and good, but I loved to just let my mind go and soar the endless reaches of space, where lines crossed to create points with no dimensions at all and parallel lines intersected in infinity. I started to think a lot about infinity, and what it was like there, and how all the postulates and theorems and principles were true across all the universe. I lay in my bed at night, Daisy Mae's head on my feet, and looked up into the darkness and allowed my mind to go wherever it wanted to go. Sometimes when I did that I actually felt like I was flying, soaring into the night sky over Coalwood and through the dark valleys and mountain hollows that marched away in the moonlight. One night, when I was having one of these visions, I had the startling revelation that plane geometry was, in fact, a message from God. My mind closed down and I came immediately back to my bed, my room coalescing around me, my desk and chair, my little chest of drawers, the books and model airplanes suddenly so terribly real. Daisy Mae stirred and I knew I was safe in my room, where I felt the safest of anywhere, but I was still trembling with fear. I lay there, unsleeping, waiting for the idea to leave me, but it wouldn't. All the next day and the next, it kept batting around in my head. I decided I had better see Reverend Lanier about it.

Reverend Lanier greeted me warily in his study. He had successfully survived his little sermon that had resulted in Cape Coalwood, but it had apparently been a close call. He told me that Mr. Van Dyke himself had suggested that perhaps the Reverend might care to review Proverbs 17:19. Reverend Lanier had and concluded that, while Mr. Van Dyke's theology was flawed, his message was clear. Reverend Lanier would take great care with his future messages from the pulpit.

Unfazed, I presented my revelation that in the principles and theorems and axioms of plane geometry—these truths that stayed true across the universe—God had sent us a message. The Reverend wasn't buying it. "You're talking about arithmetic, Sonny," he said and tapped the Bible. "All of God's words are here, in the Good Book."

I tried to talk about it some more with him, but he just kept tapping the Bible. My next stop was the Reverend Richard. Little and I walked down the narrow aisle of his tiny church toward the altar while I explained. He seemed to bow under the weight of what I was saying. "Gawdalmighty," he breathed. "Can't be nothin' but God's plan." He grabbed a Bible from behind his pulpit and plumped down on one of the crude wooden pews. I sat beside him while he opened the book and closed it and then opened it again. "The Word is the Word, Sonny," he said, running his finger along a random passage. "But the Number is God's too. Got to be." He scratched his chin, his eyes lifted to the plain wooden cross nailed to the wall by the choir box. "I can't cipher it." He looked at me. "Do you think you can?"

I shrugged. "Not me. I just want to know how to build a rocket."

"Oh, if that's all you want, pray on it and God will provide," he said. "I'll help you if you promise me somethin'. When you build your rocket and it goes off way high in the air, people may say let's give Sonny glory for it. Don't you take none of it." He nodded toward the cross. "All the glory in the world belongs right there."

I looked at the cross and then bowed my head, suddenly afraid that God might punish me for poking around in His business. "Yessir," I gulped.

"Don't be puttin' on any airs, now, gettin' prideful and all."

"No, sir," I said in a small voice, as small as I felt.

Little laughed, a kind of slow *heh-heh-heh*. "Boy, don't you be frettin'. God is love, don't you know that? He ain't never gonna hurt you. He's got plans for you, all you boys."

I nodded dumbly. "Then go on witch'a," he said. "I got some prayin' to do. Boy in Coalwood findin' the Word of God in his plane geometry book. Yes, sir. I got a lot of prayin' to do about *that*."

ONE morning, Dad plunked bread slices in the old toaster that sat on the counter and pushed the mechanism down and then went to the stove to pour coffee. When he came back, the handle on the toaster was still down, but nothing was happening. He discovered the heating element was gone, mainly because I had taken it to see if my plans for an electrical-ignition system would work.

Those same plans led O'Dell to borrow the heavy-duty battery from his dad's garbage truck. Roy Lee drove him and the battery to my house for the test. It worked, the toaster wire getting sufficiently hot to ignite black powder, but then we got distracted by *American Bandstand* and then Roy Lee and O'Dell drove off, leaving the battery and wire in the garage. For my dad, that meant no toast. For O'Dell's dad, the next day, that meant his garbage truck wouldn't start. There was unhappiness in both houses, and the gossip fence gleefully sang with it. It didn't take long before every missing thing in town was blamed on "those rocket boys." One day, Mom got a call from Mr. Jackson, who lived up in the New Camp part of town and fancied himself a hunter. "Elsie, would you ask Sonny if he's seen Jesse?"

Mom knew Jesse was Mr. Jackson's old hound dog. "Why on earth for, Mr. Jackson?"

"Well, I heard those old Cape Canaveral boys were shootin' monkeys off into outer space. I thought maybe your boy got around to old huntin' dawgs."

Mom suppressed her laughter. "Don't you worry, Mr. Jackson. Jesse will wander in. I don't think the boys have him."

Old Jesse did indeed meander back home soon afterward, but Mr. Jackson still always gave me an odd look whenever he saw me go by on my bike.

O'Dell and Roy Lee, looking for a way to improve Cape communications, targeted the mule barn. Mr. Carter had built it in the early 1930's for the old mules who had become too broken down to work in the mine. He refused to sell the beasts to the rendering factories, maintaining that they deserved something of a retirement for their years of loyalty. After years underground, they were too sensitive to light to be put out to pasture. The old wooden structure had been empty of mules since Mr. Carter had sold the company to the steel mill. As soon as the men from Ohio arrived, the mules were loaded up and sent away to become dog food. Mom and many of the women, I was told, lined the streets crying as the trucks holding the tired old animals went by. When we kids played around the mule barn, we peeked through its dirty, screened windows to look at the ghostly stalls and the ancient harnesses. On a table in the center of one end of the barn were also a number of ancient mine telephones. O'Dell, believing these telephones to be scrap, decided the BCMA should have them. Rather than simply ask the company for them, he instead hatched a plan that he thought would be more fun.

O'Dell and Roy Lee arrived at the mule barn on a Friday, near midnight. The next morning, while I was watching cartoons on television, the home phone rang. It was Tag Farmer, the town constable. "I think you better hightail it on down to Mr. Van Dyke's office, Sonny," he said. "You got problems."

When Tag told me what had happened, I felt like choking O'Dell. This stunt, I knew, had every chance of getting us kicked off company property again. I grabbed my bike and raced down Main Street.

Mr. Van Dyke raised his eyebrows when I ran in and screeched to a halt in front of his desk. Roy Lee and O'Dell looked up from their chairs by the wall. They looked miserable and dirty. "Well, Sonny," Mr. Van Dyke said, "I understand that your rocket club is in need of telephone equipment." His expression was stony, and he made a little church with his fingers. "So you decided to steal it from the company, eh? Oh, you boys think you're so sly, but we know what is going on in this town, far better than you may think. Isn't that right, Tag?"

Tag Farmer, leaning against an old wooden filing cabinet in the corner, nodded. Tag was in his official company khaki uniform, a company badge in the shape of a star on his jacket. He was a young man, still in his twenties. After graduating from Big Creek, the story I got was he had spent his entire time in Korea on top of the same mountain, waiting for the Chinese to come up and kill him. It must not have been an important mountain, because they never bothered. When his tour was up, Tag came home to go into the mine. He'd made it to the bottom of the shaft, but couldn't get off the cage. Because he was a combat veteran, Captain Laird found him another job. Tag had proved to be a good constable. There wasn't much crime to deal with, but he was always on call to come help housewives move furniture, and anybody who needed a ride could depend on Tag to supply taxi service.

"Breaking and entering with the intent to commit theft. What do you call that, Tag? In legal terms?"

Tag shrugged. "Reckon it be a felony, Mr. Van Dyke." Roy Lee and O'Dell hung their heads. My knees nearly buckled.

Mr. Van Dyke leaned back in his chair, the springs creaking discordantly. "A felony, my! Does that mean jail, Tag?"

"I'm afraid so, sir."

Instead of a future with the von Braun team, it looked as if my future was going to be behind bars. I considered throwing myself on my knees and begging for mercy. O'Dell gulped so loud I heard him. Roy Lee kept a stoic silence. Tag shifted his weight from foot to foot and then said, "Mr. Van Dyke, could we talk about this? I mean, maybe this don't need to go over to the court in Welch."

Mr. Van Dyke shrugged. "Well, if you insist, Tag, although it seems to me we'd be hard-pressed to go against the law."

Tag indicated the door. "Why don't you boys go outside and have a seat? I'll come get you in a minute. Go on, now."

We filed out and sat in the office outside. The typewriter on the desk was covered and the desk was uncluttered. Mr. Van Dyke was without a secretary again. The latest one, come down from Ohio, had spent the better part of a month in Jake's room before she got the ax. As a result, Jake had been ordered never to go out with a company secretary again, and Mr. Van Dyke's wife had let Coalwood know that the next time one was hired, she'd do the picking. "Wait'll Jake Mosby sees the spinster I choose," Mrs. Van Dyke told the fence. "And wait'll Mr. Van Dyke gets a load of her too," the fence had answered back, deliciously.

Roy Lee sat silently, glaring at O'Dell while the boy whispered furiously in my ear about what had happened. Around midnight they had arrived at the mule barn, finding a rusty padlock on the back door, knocked loose with a whack from a hammer out of Roy Lee's car. They pushed ahead cautiously, O'Dell holding the flashlight and flashing it around the empty stalls. He said it smelled like the air was a hundred years old in there. Then the rotten old floor collapsed, and the boys plummeted into the basement. Bats chattered and spun around the rafters, sailing through a cracked window into the cold night outside, and then it was quiet again. Trapped, they spent the night in old mule dung until Tag found them.

Roy Lee finally spoke. "I hate you," he said to O'Dell. He looked at me. "I hate you too." Then he lapsed into silence again.

Tag came and got us, and we went back inside and stood, our heads lowered, while Mr. Van Dyke pondered us. "How much do you think that old telephone equipment's worth?" he said finally.

We had no idea, we mumbled.

His hand moved to a big black metal calculator. He tapped some keys and pulled the lever on it and then inspected the resulting paper strip. "All right. Here's what I'm going to do, boys. If you want telephone equipment, you can have the lot for twenty-five dollars plus two dollars for the busted padlock plus another ten dollars for failure to notify the company of your plan to enter the barn. We're going to make this a business deal, gentlemen, so that your dubious records won't be further besmirched. Tag has taken up for you, although God knows why, said you boys aren't usually the ones who soap up his car or hit the sides of the doors in your cars as you drive by old people walking on the street. In short, he has asked me to display mercy toward you, even though my instinct is to take advantage of the moment and see not another rocket fired off in this town. So, is it to be business or criminal proceedings?"

"What about our parents?" I asked cautiously.

Mr. Van Dyke's eyes widened in theatrical astonishment. "I'd never recount the contents of a business deal to another party!"

And so the BCMA went into business with Mr. Van Dyke. We had a year to pay off the thirty-seven dollars, and though I had no idea how we would get the money, at least we were still building rockets. "I know how to get money, a lot of it," O'Dell said afterward, outside on the street. His face fairly glowed. "Cast iron." He laid his finger beside his nose, a sign of stealth, or perhaps deceit. "Next summer. Cast iron."

"Count me out," Roy Lee said.

OF all my subjects in the eleventh grade, chemistry was my favorite, because Miss Riley was our teacher. She was strict with us, not ever allowing anybody to get her off the topic even once, but she still had an impish humor that she often used to keep us alert, along with such an obvious love for her subject that we all paid attention. Our advanced curriculum put us into the periodic table during the first week. By the second, we were working at balancing chemical equations. If we didn't understand something, we were expected to say so and then she'd go back and patiently cover the ground again. If we didn't ask a question, she assumed we understood the material and kept going. I had at least an hour of chemistry homework to do every night. That was on top of the three hours from the other subjects.

Even though Big Creek gave Miss Riley little in the way of lab equipment to demonstrate what was in our new chemistry book, she was inventive. One day, she led us outside to the football field. The field had deteriorated during our fall of suspension. Its grass was ragged and brown, and the chalked yard lines had faded to a pale yellow. Even the stands and press box seemed to be sagging. Miss Riley poured on the ground a small amount of white powder from each of two little paper bags she had carried with her and then mixed them with a wooden spoon. I was standing beside Dorothy. To my surprise, she moved closer and took my arm and pressed her breasts against me for a few seconds before walking around to a different place to get a better look at Miss Riley. I looked up to find Roy Lee frowning at me. I sheepishly grinned.

"This is a mixture of potassium chlorate and sugar," Miss Riley said. "What we're going to see now is a demonstration of rapid oxidation. Quentin, tell us the difference between slow and rapid oxidation."

Of course, Quentin knew our homework cold. "When oxygen combines with an element over an extended period of time, the result is slow oxidation, rust being a good example of it," he said confidently. "But when oxygen combines with something rapidly, energy is released in the form of light and heat."

"Thank you, Quentin. This mixture of potassium chlorate and sugar will demonstrate rapid oxidation." Miss Riley struck a match and dropped it onto the little pyramid of powder. Instantly, a hot greenish flame erupted with a loud hiss. The BCMA looked at one another. We didn't have to say what we were all thinking. *Rocket fuel.*

After class, I went up to Miss Riley's desk and pointed at the little sack of potassium chlorate. "Can I have what was leftover?" I asked. I told her about the BCMA, just in case she hadn't heard about it. "We've built a range—Cape Coalwood—and we're starting to get some altitude. But we need a better fuel."

"Have you thought any more about entering the science fair? I'm still in charge of the committee."

"I don't think we're ready," I said honestly. "We're still trying to figure things out. It would help if we had a book."

"A book." She cocked her head, thinking. "No. I can't say I've ever seen a book on how to build a rocket. I'll look around though."

"Would you? That would be great. In the meantime . . ." I pointed at the sack.

She shook her head. "Sorry, it's all I've got. Anyway, potassium chlorate is unstable under heat and pressure. It's too dangerous for rocket fuel. What do your parents think about the BCMA?"

"My mother said just don't blow myself up."

She laughed and then seemed to ponder me as if I were some sort of puzzle. "Why do you build rockets?"

She was easy to talk to, almost like a friend. "I guess I just want to be a part of it—going into space," I told her. "Every time they launch something down at Cape Canaveral, it's like . . . I just want to help out somehow. But I can't. If I build my own rockets . . ." I stopped, not certain I was making sense.

She helped out. "If you build your own, you're part of it. I can see that. For me, it's the same with poetry. Sometimes I have to write some of my own—it's poor, I know that—but it allows me to make a connection with the poets I admire. Do you understand?"

"I think so," I told her. No teacher had ever confided in me about anything to do with her personal life the way Miss Riley had just done, almost as if I were her equal.

She kept smiling at me, and I felt at that moment like I was the most important person to her in the world. "Let me give you some advice," she said. "Don't blow yourself up. I think I want to keep you in my class. Okay?"

"Okay! I mean, yes, ma'am."

Quentin was waiting for me in the hall. "What did she say?" he asked.

"She won't let us have the potassium chlorate. She said it was too dangerous."

He clapped me on the shoulder. "That's okay. Potassium nitrate has much the same property and exactly the same number of oxygen atoms as potassium chlorate. Mix saltpeter and sugar and we should get the same reaction we just saw."

Quentin put down his briefcase and hauled out his chemistry

text. He found the equation. "Potassium nitrate. KNO_3. The same as potassium chlorate except it has a potassium atom instead of a chlorine one." He put a piece of paper against a locker and scribbled down the formula. "I think if we mix it with sugar and add heat we'll get three parts oxygen and two parts carbon dioxide along with some other byproducts. In other words, lots of good expanding gases. It should be an excellent propellant."

Quentin looked to be right. "I'll test it tonight," I promised.

Back home, I headed for the basement after a brief raid on Mom's kitchen cupboard. I took a tablespoon of sugar and the same of saltpeter, stirred them in a coffee cup with a wooden spoon, opened the door to our coal-fired hot-water heater, and tossed it in. I was gratified by the eruption of hot flame, just like Miss Riley's experiment, except mine was pink rather than green. The sound and intensity and time of the burn seemed to exceed the best of my black-powder combinations. I whipped up some more mixtures, experimenting with the percentages.

Mom was outside leaning on the fence, gossiping with Mrs. Sharitz, when suddenly our chimney erupted with smoke and sparks like a small volcano. Both came running down the basement steps just as I threw in another cup of mixture. I clanged the heater door shut and gave them my best innocent grin. "Hi, Mom, Mrs. Sharitz."

"See, Elsie? I told you Sonny was home from school," Mrs. Sharitz said.

"It would be nice to find that out without smoke signals," Mom growled.

I showed both of them what I was doing, how I mixed the propellant, how I stood back when I threw a little of it in the heater. I demonstrated, and Mrs. Shartiz whooped excitedly at the flash of pink sparks. "How pretty!"

Mom was dubious. "Okay, Sonny, let's go over this one more time. Don't blow yourself up. Got it?"

I put on my most honest face. "Yes, ma'am. I got it."

That night, I started to load a casement with a sugar-and-potassium-nitrate mix. It was too granular to attempt to put a spindle hole down the middle, so I poured it in and tapped on the casement to settle the grain as best I could. Dad caught me at it as he came in from work. "What now, little man?"

"New propellant, sir."

"If it went off, how high would this house come off its foundation?"

"Only a foot or two," I said.

"Attaboy," he said and then kept going. Startled, I turned and watched him go up the steps. *Attaboy?*

The following Saturday the BCMA gathered and went down to the Cape for a test. This launch wasn't advertised, because we had no idea how our new propellant of saltpeter and sugar would work. *Auk IX* took off with a satisfying hiss, but it quickly died and fell with a *plop* not more than a hundred feet from the pad. We recovered it and carried it back to the blockhouse to consider it. When I tapped it, a little debris fell out. Most of the propellant had burned. Sherman sniffed at it. "It smells like candy," he said.

"Rocket candy!" O'Dell chimed, and so coined our new term for the propellant.

"It seems to produce an ample exhaust, but it burns too rapidly," Quentin said. "A loose mix in the casement may not be adequate. What we need to do is somehow pack more of it inside."

"I can try wetting it with the postage-stamp glue in the next batch," I proposed.

"Sugar's awfully soluble," Quentin said, biting his thin lip. "It may retain the moisture for a very long time. You may try it, Sonny, but the proof will be here on the range, of course."

"Of course," I said back, pleased that our discussion sounded so scientific and professional.

"You guys don't have the foggiest idea what the hell you're talking about, do you?" Roy Lee asked.

Despite his statements to the contrary after the mule-barn incident, Roy Lee was still with us. Quentin scowled at him, but I laughed at Roy Lee's insight. He was right.

We launched again the following weekend. I had wet the potassium-nitrate-and-sugar mix and packed it inside a standard casement. A new member of the BCMA joined us. His name was Billy, a boy in our class who lived up Snakeroot Hollow. Other boys occasionally expressed an interest in joining us, but Billy was the first one who took it upon himself to persist in asking. I was glad to have him. Billy was a good runner, which, considering the range we hoped to attain, I thought we might need to help us find our rockets. He was also smart, smarter than me if his grades were any evidence. Billy's dad had been cut off in 1957, but had stuck around by claiming an old shack above where the colored people lived in Snakeroot. After his attendance at a BCMA meeting at our house, Mom took one look at what Billy had to wear and stopped him at the door. She drew him aside and then took him to my closet and threw open the door. Billy staggered to Roy Lee's car, weighted down with pants and shirts.

Auk X sat on the pad and fizzled, producing some white smoke and just enough thrust to rock it gently on its fins. We inspected it afterward, a dark, thick liquid like caramel oozing from it. "I cured it all week and it was still wet," I told the others.

Quentin shook his head. "I warned you. Sugar's too soluble."

Auk XI, which had rocket candy inside that I had not moistened, leapt off the pad with a satisfying hiss, but then exploded, steel fragments whistling overhead while we hit the dirt inside the blockhouse. We crept outside and stood around the pad. "My speculation is the propellant collapsed," Quentin said.

The steel casement was turned back like a banana peel. Quentin expanded on his theory. "When the rocket took off, the propellant was so loose it just fell inward. Too much of it burned at once."

"The nozzle was probably clogged too," Billy said, which was a decent observation for his first time on the range.

We went back and looked at the first rocket. The batter that dripped out had hardened. I dug at it with a stick. "There's no way this could fall inward," I said.

"But that's been melted," Sherman observed. "I wonder if it would still burn?"

To find out, we took a chunk of it to the pad and lit it. It sputtered and then burst into flame. Sherman said what we were all thinking. "What if we melted rocket candy before loading it into the casement?"

For the first time since we began building our rockets, I hesitated. "I don't know, boys," I said. "That sounds like a prescription for getting our heads blown off."

The others stood around me, looking concerned and thoughtful. "If we were very careful . . ." Billy began.

"Melted just a little bit at a time," Sherman added.

"Look, it's me who would have to do it," I said. "And I think it's just going to blow up in my face."

"We'll help you," Roy Lee said.

"I'll build us protective masks with shields and everything," O'Dell said, his eyes wide with the concept of it.

"No," I said. "It would be crazy."

We stood around in a circle, kicking at the slack. "I still say we do it," Roy Lee said quietly.

"What do you think, Quentin?" I asked.

Quentin shrugged. "This one's your call, Sonny. It is a step in the unknown, I'll warrant, but . . . damn. It would be a fantastic propellant, I'm sure of it!"

One night the following week, Roy Lee, Sherman, and I visited Jake's rooftop telescope. NASA had launched the little thirty-eight-pound *Pioneer 1* to the moon. It was America's first attempt to reach the moon, and we were excited about it. We knew we had no chance of seeing such a tiny object, but we just felt closer to it up on that roof. *Pioneer 1* arced through space until, sixty thousand miles out, not quite one-quarter of the way, it lost momentum and dropped back, burning up in the Earth's atmosphere.

The newspapers called *Pioneer 1* a failure, but it wasn't, not for us coal miners' sons on top of the Coalwood Club House. When Jake went down the ladder to his room, we stayed on the roof, talking about the moon and what it might be like, and occasionally peeking at it through the telescope just in case something about it had changed.

In fact, it had already changed because we had gone to it in our minds. We had flown the little spacecraft beyond its physical capabilities, zipped past jagged mountains and over the gouges and tears of primordial bombardment, admired all the moon's craters, its *mares,* and its mountains. Someday, I was convinced, we would go there. Not just mankind, but *us,* the boys on that roof. If only we could learn enough and were brave enough. That's why I decided, up there on that roof, that we would melt saltpeter and sugar.

11

ROCKET CANDY

Auks XII–XIII

ON SATURDAY MORNING, I began stringing extension cords from the basement up the steps and out into the backyard and over the fence and into the back alley to the other side of the garage, where there was an old picnic table. O'Dell had assembled protective shields for us—baseball caps with squares of clear plastic, rescued out of some garbage can, taped to the bills so they hung down in front of our faces. I put on an old Navy pea coat (it had belonged to my Uncle Joe when he was in the Navy during World War II) and winter gloves. The hot plate and pot were from one of Mom's kitchen cabinets. I didn't figure she'd miss them before I put them back.

The hot plate glowed a bright orange. The other boys stood back as I sprinkled saltpeter into the pot. I prayed it wouldn't blow up in my face. A few drops of liquid formed and then boiled off. Encouraged, I took a tablespoon of saltpeter and dumped it in and then stirred the pot with a wooden spoon also "borrowed" from the kitchen. A clear puddle of liquid formed, and O'Dell, also wearing a ball-cap shield, coat, and gloves, poured in more until there was an inch of liquid steaming in the bottom of the pot. "Now the sugar," I croaked. Fear made my throat dry.

O'Dell leaned back at an acute angle and shoveled a little sugar in. Nothing happened except the granules immediately dissolved and a sweet smell, not unlike vanilla fudge, began to wisp up from the pot. Encouraged, he poured more sugar in. I kept stirring until the mix turned viscous and milky.

"I'll be gawddamned," Roy Lee breathed in relief. "It didn't blow up."

"Don't cuss," I scolded, sweat trickling into my eyes. "Pray."

Sherman put on his cap shield and put an inverted *Auk* casement on the table. "We need a funnel," he said.

Roy Lee was rummaging through the cupboard in the kitchen when my mother found him. "Hi, Mrs. Hickam," he said, grinning sheepishly. "Sonny needs a funnel."

She eyed him suspiciously. "The only one I know of is the one in the garage that his dad uses to change oil in the Buick. What does he want it for?" Then she looked past him to the extension cords. "What are you boys up to now?"

"Um, melting rocket fuel."

Mom held the funnel as she came around the garage. "It smells like fudge," she said when she caught a whiff of the mix. At her appearance, we all froze. "Well, go ahead," she sighed. "Do what you're going to do."

Roy Lee gently took the funnel from Mom's hand and inserted it into the *Auk* casement. I lifted the pot and tipped it carefully, the rocket candy pouring out. But it hung up in the funnel, backing up and almost overflowing. Mom hurried into the garage, coming back with a long straw torn from an old broom. "Here," she said, jabbing at the mixture in the funnel.

"Mom!" I protested, and Roy Lee tugged her arm back. If the stuff caught fire, she didn't even have one of our pitiful little shields.

Roy Lee moved Mom back to the garage, and O'Dell took over the straw jabbing. It worked, and the slurry began to flow into the casement. "A glass rod would be better," Sherman said for future reference.

The casement was only half full, so I put the pot back on the hot plate to mix up some more. That was a mistake. A thin layer of melted mix had dried on the bottom of the pan. With a *whoosh,* the stuff erupted.

"Whoa!" We all fell back. The pot went flying. A puff of vaporous spume rose from the back alley like an Indian smoke signal.

A group of men going to work stopped to watch. "Hey, Elsie," one of them called. "You teaching those boys how to cook?"

Guffawing, several of them strolled in closer, carrying their lunch buckets, their helmets set back on their heads, pants tucked in hard-toe boots. I recognized them as the crew of men out of Anawalt that Dad had brought in for a one-time job, the demolition of a particularly stubborn slab of rock in a new part of the mine. They were sharing a house up in the New Camp part of town. "You boys are stupid, ain'tcha?" one of them said, his big chaw shifting in his cheeks.

Mom's eyes narrowed. "Get on to work. These boys aren't stupid. They're scientists. I said *get!*"

The men trudged on, laughing, and we were left alone with Mom. Every muscle in my body said *run,* and the other boys seemed halfway at a trot already, although they were actually quite still. Mom picked up her blackened pot and contemplated it. "I think if you wash the pot after you mix this stuff, it won't explode." She tapped our plastic face shields, tugged at our coats, inspected our gloves. I started to explain everything, but she held up her hand. "Move that table farther from the garage. You wouldn't want to burn up your dad's Buick." She looked at me. "You'll buy me a new pot."

"Yes, ma'am."

She looked at us, one by one. "I'm getting tired of saying this: Don't blow yourself up!"

"We'll write everything down like a recipe," I said, trying to assure her. "Wash the pot, clean up everything before we melt another batch."

"For sure. Yes, ma'am," the others mumbled, shifting back and forth on their feet. They still weren't sure they were going to get out of this without big trouble from Mom.

"Will this make your rockets fly better?" she asked.

We looked at one another. We were scared truthful. "Maybe" was all I could say.

LATE one evening, when usually he was dozing in his easy chair in front of the television set, my father opened the door of my room and walked inside. He caught me idling, doodling the design of a rocket that could fly to the moon. My homework, a stack of it, awaited. "I hear from your mother that you're thinking about being an engineer," he said.

"I guess I don't know what an engineer does," I said. "All I know is I want to work on rockets," I added, just in case we were having an argument that wasn't apparent yet.

"There's a lot more to engineering than rockets," he said gruffly. He smoothed his voice, as if catching himself. He picked up my drawing and looked it over. "If you're thinking about being an engineer, you need to see what one does for a living." He put the drawing back on my desk and looked around my room, one of the first times I think he had ever done that. I had a couple of *Auk* casements on my dresser, which he stared at for a moment. "All this rocket stuff they're doing down at Cape Canaveral," he said, "is just burning up taxpayer's money to scare the Russians. A real engineer builds things to make money for his company."

"Yessir." I thought by agreeing with him, maybe he'd go away.

"I'm going to show you firsthand what an engineer does," he said.

Then he told me his plan. It was the most remarkable thing. I just gaped at him. "Are you sure?" I puffed up a bit. This was something he'd never asked Jim to do.

"I'm sure," he said. "It's about time you see what this town's all about."

BASIL was at the next launch with his notebook pad, writing furiously. The response from his articles had been so great he had decided to make us a regular feature. Watching us were about fifty spectators, attracted by his articles and our notices. I hoped we wouldn't disappoint them. *Auk XII*, on the pad, was built to the design we had more or less standardized, but how it would perform loaded with melted rocket candy was a complete unknown. I feared an explosion, and Quentin was certain of it. "It'll take at least three rockets loaded in this manner," he predicted, "before we get the right combination." Sherman went over to the road to make sure people knew they needed to stay back and it would be best if they got behind their cars and trucks. Buck and some of the football team showed up too. They stood apart, quietly sullen.

Except for the football boys, our audience was festive. "Go, Big Creek," some people called out, just as if we were the football team. Then, when we ran up our BCMA flag, they began to sing the school fight song. *"On, on, green and white, we are right for the fight tonight! Hold that ball and hit that line, every Big Creek star will shine. . . ."*

I had never known what it felt like to be on the receiving end of that song. I liked it. A football boy yelled something derisive, but the people kept singing. Afterward some young ladies chanted like cheerleaders, "Go, rocket boys, go!" Disgusted, Buck and the other team members got into their cars and left.

This was also our first launch with an electrical-ignition system. I touched a wire to a car battery (an old one O'Dell got for free from a War junkyard), and *Auk XII* shot off the pad and leaned down-range. Quentin ran outside the bunker and fumbled with a new invention he called a "theodolite." It was a broomstick with an upside-down protractor attached on one end and a wooden straightedge on the opposing side that rotated around a nail. He jammed the stick in the slack and went down on his knees and squinted along the straightedge at the rising rocket, smoke squirming from its tail, white against the brilliant blue of the cloudless sky. At the height of the *Auk*'s climb, Quentin looked at the angle the ruler made with the protractor and called it out, then snatched a pencil stuck behind his ear and wrote it down on a scrap of paper. If his theodolite worked, trigonometry would give us the altitude of our rocket.

Auk XII's exhaust trail was still a fast stream when the rocket faltered and began to fall. It continued to smoke vigorously even after it struck the slack. While our audience cheered, we ran after our rocket and watched the last of its sputtering rocket candy burn up. I immediately saw the reason our rocket had lost its thrust. "The nozzle's gone," I told the others. "It must have blown out."

We looked closer. The weld was intact. The center of the nozzle was simply eaten away. Quentin came stepping up to us. "Three hundred and forty-eight," he said, finishing his count by bringing both his feet together at the final step. "I'm figuring about two point seven-five feet per step. That would be"—he made a quick mental calculation—"nine hundred and fifty-seven feet." Jake's trig book was under his arm. He ran his finger down the functions in the back. "Let's see, the tangent of forty degrees is about point eight-four. Call it point eight. Multiply that by nine hundred and sixty . . ."

We waited anxiously while Quentin worked it out in his head. It didn't take long. "Seven hundred and sixty feet!"

O'Dell whooped and did a little jig on the slack.

Auk XIII jumped from the pad in a similar blurred frenzy to its predecessor. Rocket candy was definitely hot stuff. The rocket leaned over, puffed a big cloud of smoke, and sped off into the sky. When it fell back, it disappeared into a dense thicket of trees. We heard it hit branches as it fell, a big oak tree waving its golden leaves at us as if signaling, *Come and get it. Rocket over here.* O'Dell knocked over Quentin's theodolite in his excitement, so we didn't get an altitude estimate, but it was obvious it hadn't gone as high as *Auk XII*. When we found the rocket, the nozzle was completely worn through. "Maybe it just can't take the heat," Billy said.

I studied the nozzle. "You know what? It looks to me like it's corroded," I said.

"Rapid oxidation!" Quentin said, snapping his fingers. "Sonny, my boy, how quick you are! Of course! I should have seen it myself! Just like in Miss Riley's class. Heat combined with a steady flow of excessive oxygen—it makes sense. What we need, gentlemen, is a material capable of withstanding heat *and* oxidation."

When we came down off the mountain with our rocket, all the observers were gone, but Buck and the football team were back. They were at our blockhouse, tearing it apart, board by board, using tire irons.

We roared and ran at them.

"Come on, little sister morons!" Buck screamed, red in his face.

We were no match for them, but we had to do something. I picked up a rock, and the other boys did the same. We let loose a barrage, missing for the most part but making them dodge. They charged us, and we knew we were doomed. Then we heard a car horn and Tag Farmer drove his old Mercury out on the slack. While we, football boys and rocket boys, all froze in place, Tag leisurely got out and pushed his constable's cap to the back of his head. "So, what's going on, boys?" he drawled.

"Nothing," I said. I wasn't about to turn Buck in. Boys in Coalwood just didn't do that. "We were just cleaning up the range."

Tag nodded toward Buck and the others, still standing with their fists balled and tire irons in their hands. "They helping you?"

"Yessir."

Tag strolled over to the blockhouse, musing over the planks that had been ripped out. "Buck?" he called softly.

Buck meekly went to the constable. "Yessir?"

"You a carpenter?"

"No—no, sir."

"It might be time to learn. Looks like some boards came loose on this blockhouse."

"Yessir."

"You going to take care of this?"

"Yessir."

Tag nodded. Buck stooped and picked up the boards. I came over and handed him a hammer from the tool chest we always carried with us, and he got busy. Tag chuckled and stayed around until everybody left.

On Sunday morning, I pretended to oversleep, part of Dad's plan to avoid trouble, and Mom flung open the door to my room. "Get up or you're going to miss Sunday school!"

I was about to fib to her, but I had decided it was okay since it was for her good. "I'm pretty worn out, doing so much homework and all. Would you mind if I skipped today? Just this once?"

She turned and went out of the room. "If you want to be a heathen, who am I to stop you?" She went bawling after Jim to get out of the bathroom and drive her to church. He answered back that he had only been in there for a couple of minutes. I figured it had been at least an hour.

After Jim and Mom left for church, I walked up the path to the tipple, where Dad waited for me. I was almost shaking with excitement. I'd lived in Coalwood my whole life, but had never been where Dad was going to take me. I was going in the mine! And the fact that he'd asked me, not Jim, had grown even more important through the week whenever I thought about it. He eyed me carefully when I came into his office. "You didn't say a word about this to your mother, did you?"

"No, sir!" I said it loud and proud.

"Right. We'll get you washed up afterward and she won't ever know the difference."

That part of Dad's plan I had my doubts about, but I happily went along with it. He knew Mom better than I did, after all. "Come over here," he beckoned, spreading a map of the mine on a table. He pointed at a winding black streak that ran across it. "That's the Number Four Pocahontas Seam, the finest and purest soft coal in the world. These lines I've drawn represent the tunnels we've driven through it since the mine has been operational." He opened a drawer and brought out another drawing. "This is the side view of a typical seam. The coal is overlaid by a hard shale called draw rock. Underneath is what we call jack rock. Engineers have to know how to hold up the draw rock to keep it from falling and how to move the jack rock out of the way.

"Doing the engineering in a mine takes a lot of experience and careful calculations," Dad continued while searching my eyes. I think he was looking for a glimmer of understanding. "Men who work under those roofs depend on the mining engineers to do it right the first time. It's not like your rocket men—those crazy German scientists—just throwing something up to see if it will work."

I resisted the urge to answer that charge, and Dad lectured on. The coal company used the block system, he said, each block being seventy-five feet by ninety feet. Entries were driven through them in sets of four. After that, the blocks were taken out by continuous-mining machines, one by one, until each of them was only about fifteen feet square. Those remaining blocks were called pillars, which were also eventually mined. During all of it, roof bolts and posts and cribs all had to be calculated and set to hold the roof up.

Then Dad went off into his favorite subject: ventilating air through the mine. "If the air stops circulating, methane will seep out of the coal and build up," Dad said. "One spark and the whole mine could explode. To keep that from happening, we use a pressure system. Fans raise the pressure in the mine to a little greater than the surface pressure. The methane is blown out through the vents."

"You designed that?" I asked.

"I did a good portion of it," he answered, glancing down at his drawings.

I was confused on this point. "So you're an engineer?"

He toyed with a slide rule. "No. An engineer has a degree."

I decided to use some of Mr. Hartsfield's deductive reasoning. "Jake Mosby's an engineer," I said.

"That's right."

"You know a lot more about coal mining than he does."

"That's true."

I shrugged. "Then you're an engineer, right?"

Dad shook his head. "Sonny, you have to have a diploma from a college to be an engineer. I don't have one. That means I can never be an engineer." He looked at me speculatively. "But you could."

Not knowing what to say, I didn't reply, but kept studying the drawings. "This is interesting," I said, and meant it.

Dad led me to the bathhouse and opened his locker and handed me a one-piece coverall, hard-toe boots, a white foreman's helmet, and a leather utility belt. When I joined him at the man-hoist, he showed me how to clip a lamp battery pack onto my belt and the lamp on my helmet. With the lamp attached, the helmet felt heavy. I moved it around until it felt comfortable. He appraised me and readjusted the helmet, then my belt until the buckle was squared in the front and the battery hung exactly off my right hip. I felt like a soldier under inspection. "Now you look like a mine foreman," he said after another critical assessment. "Let's go."

The attendant swung the gate aside, and for the first time in my life I stepped onto the wooden-plank platform of the lift. I thought of all the times when I was a small child and had watched the miners descend into the darkness. Now it was my turn! I could feel my heart speed up. The boards in the floor were set apart enough that I could see between them. There was nothing beneath us but a dark chasm. I had a momentary twinge of fear that we were going to fall. The bell rang three times, announcing that we were about to be let down. I took a deep, ragged breath. The man-hoist winch began to creak and the lift dropped quickly, my stomach lifting up around my throat. I grabbed Dad's arm, then quickly let go in embarrassment. He said nothing, and I watched the solid rock of the shaft slip past. Men had hand-dug the shaft, but I couldn't imagine how. It had taken me and the boys all day just to dig out a little place for our blockhouse at Cape Coalwood.

Through the gaps in the floor, I started to see lights far below. Above us, the square of light at the top of the shaft had shrunk to a tiny twinkling star. We were being swallowed by the earth, and I hadn't decided yet whether I liked that. I remembered that Tag had frozen at the bottom of the shaft, refusing to get off the lift. Now I understood his fear very well.

When we neared the bottom, the lift slowed, jerked a few times, and then settled level with a rock platform. I switched on my helmet light. There were miners waiting on the platform. Mr. Dubonnet was among them. He looked at me with surprise. "A new hire, Homer? He'll need to join the union."

"Sonny's thinking about becoming a mining engineer," Dad snapped. "A company man."

"Well, well," Mr. Dubonnet replied with a noticeable lack of enthusiasm. "Now, wouldn't that be something?"

Solid gray walls surrounded us. I felt almost as if I were on some alien planet. All the things I'd ever known that oriented me—trees, the sky, the mountains—none of them was around. The air even smelled different, like wet gunpowder. Off to the right was a set of tracks with a big yellow electric locomotive sitting on it, some cars behind. I could see a connecting tunnel to the left, the blue haze of fluorescent lights showing through the window in a concrete block building. Hot white flashes and rapid hisses within indicated arc welding. Dad saw me looking. "We put a little machine shop down here. Saves time bringing out equipment that needs repair."

"Is Mr. Bykovski in there?" I wondered.

Dad rocked on his hard-toe boots. "Ike's not a machinist any-more, Sonny. He's a loader, and a damn good one." He stepped forward. "Come on. Let's get on down the line."

Dad led me to the locomotive, stopping to talk to its operator. I recognized him—Mr. Weaver, whose son Harry was five grades ahead of me. Harry had gone in the Marine Corps, had landed in Lebanon when President Eisenhower had decided to help out over there. Mr. Weaver sat on the front with a lever to control the power to the locomotive's electric motors. "Hey, Sonny," he greeted.

"Hey, sir."

"Take us all the way to the face, Frank," Dad said.

"You got it."

Dad took me to an attached car he called a "man-trip." It was a low-slung steel car with two hard metal benches inside that faced each other. We crawled inside the man-trip and sat side by side, facing forward. Dad slapped the top to let Mr. Weaver know we were ready to go. The man-trip lurched and we were off, plunging down what seemed an endless black tunnel. Dad said we were on the "main line." For twenty minutes, the rails clacked beneath us, the posts holding up the rock roof blurring past like a subterranean forest of gray tree trunks. On straightaways, the locomotive roared as we flew down the track, the man-trip rattling and shaking. I could smell the hot odor of the locomotive's electric motor. Before we got to a curve, Mr. Weaver applied the brakes, and the steel wheels of the locomotive and our man-trip squealed like a thou-sand tortured pigs. I held the steel seat with my hands between my legs so I wouldn't fall over when we took the curves.

As we sped along, I occasionally saw the flash of miners' lamps down branching tunnels, but it was too dark to see what they were doing. At my question about them, Dad said they were "dusting":

spreading rock dust around to hold down the explosive mixture of
coal dust and air. It registered on me after a time that the mine was
not the cold, dank, ugly place I'd always imagined it to be. The air
was cool and dry, and when we stopped at a switch to let a line of
low coal cars go past us, I peered down the tunnel and the mica in
the rock wall sparkled like diamonds.

I remembered then that Dad had once brought some mica crystals home with him and left them on the kitchen table for Mom,
along with a card that said: *You always wanted diamonds, but these are
the best I can do. I wish they were real.* The next morning, waiting for
him on the table was Mom's note of reply: *I never wanted diamonds. I
only wanted a little of your time. That's still all I want.* But she didn't
throw Dad's diamonds away. I knew. I had come across them and
the notes while looking for some writing paper in her desk.

When the man-trip stopped, Dad jumped out. "We've got an
operation going at the face today," he said. "I want you to see it."
When I climbed out, I stood up and slammed my helmet into the
roof so hard it almost knocked me to my knees. I staggered, then
looked up to see what I had hit and saw slabs of rock, roof bolts
jammed into them every few feet. Dad ignored my trouble and
took off at a fast pace, never looking back. I took off after him, my
helmet whapping against the roof in a painful staccato. Every time
I thought I had found a rhythm to my walk, I hit my head again.
Once I hit a header so hard it knocked me off my feet. I landed on
my back, my helmet flying, saved only by the lamp cord attached to
the battery on my belt. I scrambled after it. By the time I got my
helmet back on, Dad had disappeared around a turn. I could see
the jumping reflections of his lamp on a far wall. I hurried after
him, my helmet still knocking against the roof. I was developing a
powerful crick in my neck. Pretty soon, he was so far ahead of me I
knew I would never catch up. I was close to panic. What if I got
lost? If my lamp went out, nobody would ever find me again!

Then I heard a noise, like the mine was tearing itself apart. I felt
like running away, but where would I go? I turned a corner and I
beheld an astonishingly huge machine, spotlights bolted to its side,
tearing at a wall of coal. Dad was off to the side, watching it. He
saw me and waved me over.

"That's a continuous-mining machine!" he yelled over its roar. It looked to me more like some kind of great prehistoric animal. Dad positioned me out of the way when a shuttle car darted in, its crablike arms sweeping up the coal thrown out behind the continuous miner. When the shuttle was full, it backed off, dragging a thick electrical umbilical behind it. I realized this was the kind of vehicle Mr. Bykovski had been assigned to. I looked closer to see if it was him, but the operator was Mr. Kirk, the father of Wanda, a girl in the class behind mine. Wanda had a great voice and sang in the school choir. Mr. Kirk made a run to the track to unload his shuttle into a waiting coal car.

The noise was deafening. Dad yelled into my ear, explaining what I was seeing. The miner was working a block, he said, sloughing off the coal until all that was left was a pillar. "An engineer has to study the rock above these pillars! If weight becomes concentrated on one of them, it'll explode! The last time that happened, it tore one of our shuttles to pieces!"

I looked at Mr. Kirk's shuttle as it returned and tried to imagine the power it would take to tear such a machine apart. I wondered what had happened to the driver of the shuttle and started to ask, but Dad interrupted my train of thought. "This is real engineering work, Sonny!" he yelled, sweeping his hand around the busy workplace.

The foreman of the work party saw us and came over. The black face under the white foreman's helmet was my Uncle Robert, Mom's brother. "Homer, Sonny," he said, looking at me long and quizzically. "How's Elsie?"

"Fine, Bob, fine," Dad said distractedly.

"Does she know Sonny's down here?"

"I wouldn't take him anywhere he's not safe," Dad said. "That's all that matters."

"I wonder if his mom would agree with that," Uncle Robert replied amiably, cocking an eyebrow.

"You just let me worry about Elsie," Dad said firmly.

Dad and Uncle Robert started discussing business and I wandered away, trying to get a better angle on watching the continuous miner and the shuttle go through their choreographed dance of mining coal. Uncle Robert came and got me. "That's not a good place to stand," he said. He carried a three-foot wooden pole with him and used it to poke at the ceiling. A big, jagged rock came loose and hit the floor with a heavy *thump* right where I had been standing. I jumped and whapped my helmet once more against the roof. How my neck hurt! Uncle Robert chuckled. "A man has to be thinking every second down here, Sonny."

I positioned myself where Uncle Robert pointed, under a roof bolt, and kept watching until Dad led me back to the man-trip. As we trundled back down the main line, I was thinking about all that I had seen. I couldn't wait to tell the boys, but then I thought I couldn't do that—this was supposed to be a secret. I was worrying how I was going to get around that when Dad suddenly started to talk. "I love the mine," he said. "I love everything about it. I love getting up in the morning before sunrise and walking up the path to the tipple. I love seeing the shifts change, the men bunching up at the man-hoist, ready to go to work."

I listened, amazed, not that he was saying what he was saying, but that he would share such thoughts with *me*. I felt proud, grown-up. Dad took off his helmet and rubbed his head, scratching around it where the sides of the helmet had pressed in his hair. When he started to talk again, I focused on his every word as if they were gold coins he was dropping into my hand, one by one. "I love going to the face. I go every day even though I don't have to. That's where I see if my plan for the day is working. I see it all in my head days before, see the cut the continuous miners will take, the route of the loaders, the roof bolts going in, the places where the methane might build up and where the foreman needs to check with his safety lamp. It's all there when I arrive, just as I saw it, and I get great satisfaction from that."

I found that I was staring at him, the narrow beam of my lamp focused on his face like a spotlight on an actor. "Every day," he said, "I meet with Mr. Van Dyke and his engineers. Even though I don't have a diploma, I know more than they do, because I've been to the face and they haven't. I've ridden the man-trip down the main line, got out and walked back into the gob, and felt the air pressure on my face. I know the mine like I know a man, can sense things about it that aren't right even when everything on paper says it is. Every day there's something that needs to be done—because men will be hurt if it isn't done, or the coal the company's promised to load won't get loaded. Coal is the life blood of this country. If we fail, steel fails, and then the country fails."

The beam from his helmet lamp shone in my eyes. "There's no men in the world like miners, Sonny. They're good men, strong men. The best there is. I think no matter what you do with your life, no matter where you go or who you know, you will never know such good and strong men.

"You're my boy," he said, and then turned so his lamp shined down a side cut, the lamps of his men flashing back from the darkness as if they knew he were passing. "I was born to lead men in the profession of mining coal. Maybe you were too."

You're my boy. In the dark, I could savor the words without embarrassment.

After the man-trip got back to the bottom of the shaft, Dad led me to the lift. The cage was beneath us, the shaft a little deeper than the main level. He pushed a brass button and a bell rang and the cage came up. Dad pushed the bell twice more and we stepped aboard. A miner came out of the little cement blockhouse and, at Dad's nod, pushed the bell three times. We started up at a good clip. About a hundred feet up the shaft, the lift suddenly stopped. I looked around worriedly. The ragged cut rock surrounding us seemed to close in. All I could see above was a tiny spark of sunlight. I noticed a detail I hadn't seen before, a set of steel steps. At my question, Dad said they went all the way from the top of the shaft to the bottom. There was a gap between the cage and the steps. To get on them would require taking a step over the oblivion of the deep hole beneath us. At that prospect, my heart sped up and my hands started to sweat. "Don't fret," Dad said, perhaps sensing my unease. "They're probably just greasing the hoist."

We stood quietly for a long minute. "So what do you think about the mine?" Dad finally asked.

I knew what he wanted me to say and it was tempting to say it, but I also didn't want to lie to him. I considered my answer and settled on the kind of evasive response I'd used on my mother over the years. "I learned a lot," I said and left it at that.

I couldn't fool Dad any better than I could Mom. The difference was he didn't see any humor in it the way she usually did. "What I'm asking you is do you want to be a mining engineer?" he demanded. "If you do, I'll pay your way through school."

I carefully crafted my answer. "I'd like to be an engineer," I told him.

"A mining engineer?" he pressed.

He had me. I had no choice but to tell him the truth. "I want to go to work for Dr. von Braun, Dad."

He didn't hide his disappointment. "You should talk to Ike Bykovski about those damn Germans running around loose," he muttered.

"Sir?"

"For chrissake, don't you even know Ike's a Jew?" Dad snapped. "He'll tell you that von Braun German bastard's not worth hanging."

The lift jerked and once more we started up. I glumly watched the rock slide past. I had really messed up this time. Dad was not only mad at me, I knew I'd hurt him too. And what had he meant about Dr. von Braun and Mr. Bykovski? I blamed myself for everything. I should have never agreed to go down in the mine. I knew what Dad was likely to be getting at, and I knew I wasn't going to agree to it. So why had I gone along with it? I was a stupid kid sometimes, no doubt about it.

As we neared the surface, the cold, fresh air off the mountain blew down the shaft, giving me a shiver. The earth's surface slipped past and there, standing at the gate still in her church clothes, was my mother. A rock-dust crew stood nearby, their eyes locked on her. They shifted their gaze to Dad and me. Mom stared at what I knew must be my grimy face, my coal-mascaraed eyes, and my blackened coveralls. Then, to my utter astonishment, she burst into tears. The rock-dust crew took a step backward. Some of them took off their helmets, rubbing their heads and looking down at their feet as if embarrassed to be witnesses to her tears. Dad tried to shush her. "Stop it, Elsie, you're scaring the men," he said while unlatching the lift gate.

"Everything's fine, Mom," I said, my stomach bottoming out: We were about to have a family argument right here in front of God and everybody. I couldn't imagine anything more embarrassing.

"He's thinking about being a mining engineer," Dad said doggedly.

Mom's tears seemed to dry up instantly, as if they'd been sucked back inside her. "Over my dead body," a voice from deep inside her said.

Dad pushed me ahead. "Go take a shower," he said gruffly. He looked around at the men still ducking their heads, but watching all the same. "This is none of your business!"

The rock-dust crew moved off only a couple of feet so they could keep listening. I walked to the bathhouse, but stopped at the door. Even though I was so mortified I wanted to disappear, I wanted to hear too. Mr. Dubonnet, wearing his miner's helmet with his street clothes, stepped outside and saw what was happening. He crossed his arms and leaned against the wall, chuckling. I didn't know what he thought was so blamed funny.

"What the devil's wrong with you, Elsie?" Dad hissed, reaching to take her arm.

She pulled back. "This mine's killed you, but it's not going to kill my boys!"

"You're talking nonsense."

"Black spot—about the size of a dime," Mom said, poking her index finger at his chest. "Here, on the right side!" She poked him again—hard.

Dad huffed a near laugh, reached down for a handful of coal dust, and tossed it in the air. He took a deep breath of it. "I thrive on this stuff. It's like mother's milk to me!"

Mom watched the dust settle around him. Some of it blew in her face, sticking to her makeup, but she didn't flinch. She turned and marched inside the bathhouse where I had retreated, sending naked miners scrambling for towels. She grabbed me by my arm. "You can wash at home," she growled. As we came out, Mr. Dubonnet tipped his helmet to her, but all he got in return was a dirty look. The rock-dust crew scattered before her. Only Dad stood his ground, his helmet in his hand. He watched us pass without comment. All the way down the path from the tipple, I could feel his eyes boring into the back of my head.

12

THE MACHINISTS

Auks XIV–XV

WHEN I NEXT ventured down to my lab, I discovered Mom had taken back her kitchen hardware. All she had left me was a ruined rocket-candy pot. I had a lot of experience dealing with my mother when she got mad at me. The best approach was to throw myself immediately on her mercy. I sought her out, finding her in the kitchen. "Mom, I'm *really* sorry," I said, my head bowed. I watched her out of the tops of my eyes to see what impact my declaration had made.

She gave me a short, hard look and then stirred the pot of beans she had on the stove. "First off, you lied to me, and on a Sunday to boot," she said.

"I don't know what I was thinking," I said, allowing a little groan of regret to steal into my voice.

"Far as I can tell, you weren't thinking at all," she snapped, now furiously stirring.

"I'm sorry."

With all her heavy-duty stirring, her beans were starting to look like pudding. She stopped and threw me an apron. "You're not going to mope around the kitchen without being put to work. See those kidneys on the counter? Cook them up for the cats."

Grateful to be doing something in her presence, I tied on the apron, put the slimy kidneys in the sink, washed them, and prepared a pot for boiling. "Keep stirring these beans too," Mom said. "Don't let them stick to the bottom of the pot. I'm going into the living room and put my feet up and watch television like the Rockefellers."

"Yes, ma'am," I said, miserably. But I wasn't miserable at all. I was happy. Mom had set the terms for my forgiveness and they were pretty easy. I cooked the kidneys, my nose wrinkling at their stink, and stirred the beans, although they hardly needed it. Daisy Mae rubbed up against my leg, and outside I heard Lucifer beg for entrance. The two cats, gurgling and purring, pounced on the organ meat when I set it down for them. It smelled like hot urine, but at least the cats and my mother were pleased. I went down in the basement and opened up a can of dog food for Dandy and Poteet, petting and patronizing them to make up for the attention I had given the cat enemy upstairs. Then I went outside and scattered seeds for the birds on the picnic table and threw some old lettuce and carrots out for the rabbits. I went out on the enclosed side porch to see how Chipper was doing and fed him too. He was racing along, his truncated tail held high, in the wheel Mr. McDuff had built for him. Chipper's tail was cut short because he had caught it in the wheel and chopped about a quarter of it off. Mom had tried to tape the severed portion back on with some of Dad's green mine tape, but it wouldn't take. Chipper loved his wheel. As Mom said, he might not be getting anywhere, but at least he was getting there fast.

As I scurried past her, making a little extra noise so she'd know how hard I was working, she looked up from the television. "I washed your rocket things and put them in a paper sack in the cupboard," she said.

"Thank you, ma'am."

Mom hid a smile. "You've scraped and kowtowed enough, Sonny," she said. "Don't overdo it. And Sonny?"

"Yes, ma'am?"

Her expression was dead serious. "You ever go down in the mine again, I'll get your Ground-Daddy's old pistol out of my cedar chest and shoot you dead on the spot."

It was news to me she had a pistol in her cedar chest, but I didn't doubt it. And if it was Ground-Daddy's, I guessed it would be so big she would need both hands to hold it up and it would have a bullet the size of a hickory nut.

According to what Roy Lee heard from his mother, the fence-line telegraph had already gleefully dissected Mom and Dad's fracas at the man-hoist. Everybody was just waiting for the next chapter in the Hickam family soap opera. Embarrassed, Jim had once more lowered himself to speak to me, but only to suggest that it might be a good idea if I more or less disappeared forever. We were trudging home after getting off the school bus.

Jim was so easy. I stuck my verbal stiletto in him in just the right spot and twisted it. "I've been wondering," I said, appraising him with mock concern. "Do you think you're getting fat because you're not playing football or because you basically swallow the refrigerator every night?" Jim sputtered and threatened, but restrained himself from battering me on the street. It would have meant even more embarrassment.

Although I didn't care about Jim and had time-tested ways to handle Mom, Dad was trickier. I had disappointed him before, but it had never been in such a personal way. I kept trying to think what I should say to him, but couldn't figure out what it might be. He gave me no opportunity, in any case. For the next several weeks, he retreated to the mine, coming home after I was in bed and up and gone before I woke up. I worried about Dad, but I didn't dwell on him. Rocket-building had begun to teach me a different way to think. There were always so many things to do and to remember in rocket design and construction that I had been forced to get organized in my mind. I had learned to put all the steps I needed to put one of the *Auks* together into different categories. Then I had put all of them in the order they had to be done and of their importance. It was sort of like putting stuff into different drawers in my mind and then remembering which drawers I needed to open and when. When I told Quentin about it, he called it a "sequential approach" to problems and admired it. "It is what I have always believed to be true," he said. "Our work with rockets will change us in ways we would not have predicted. You, for instance, have actually learned an orderly way to think. When I first met you, I would not have believed that was possible."

I took Quentin's comment as a compliment. I couldn't do anything about Dad, not now, so I left his drawer closed until I could. But there was one thing—what my father had said about Mr. Bykovski on the cage—that kept bothering me no matter what else I was trying to think about. It was the drawer that stayed open. I would need help to close it.

IKE and Mary Bykovski didn't have a telephone, so I walked to their house after school. Mrs. Bykovski opened the door and frowned when she saw it was me. She was a slender woman with a pale, thin face and sunken cheeks. Her brown hair was short and straight, uneven around the edges as if maybe she had cut it herself and hadn't tried too hard to make it come out right. Mom said Mrs. Bykovski always looked a little "peaked." "The Mister's still napping," she said. "Working at the face wears him out."

I blurted out an apology for causing his job change. "It's all right, I guess, Sonny," she said in a kindlier tone. "The paycheck's bigger, anyway."

She invited me in and I sat alone on the couch in their living room while she went upstairs. The house smelled of corn bread and beans, the Coalwood staple. I saw a pair of reading glasses on a table beside the couch and also a book. Closer inspection revealed that it was a novel titled *The Fountainhead*. I didn't know it. Opposite the couch was a television set in a dark, heavy console. On top of it was a framed photograph of Esther, the Bykovski's daughter. She was in a wheelchair, her head lolling on her shoulder. Mrs. Bykovski stood on one side of her and Mr. Bykovski on the other. None of them was smiling.

After a while, Mr. Bykovski came down the stairs into the living room, yawning and stretching his suspenders over his shoulders. "Hello, young man," he smiled as his Mrs. set down a saucer and a cup of steaming tea on a little table beside his easy chair. When she asked me if I wanted something, I said, "No, ma'am," and she

disappeared into the kitchen. He noticed I was looking at the photograph on the television and said, "Maybe the day will come when
we can bring Esther home and she can go back to school with
you."

"I hope so, sir," I told him. Actually, Esther had been a disrup-
tion during the first and second grade, and I had been shamefully
glad when she left. Usually, she sat silently at her desk, staring
blankly at the teacher or with her head down on her arms. But at
other times, she would gyrate and make grunting sounds and spas-
modically sweep her books and pens to the floor. Our teachers
would wait patiently until she was through and then direct one of
us boys to pick up her things and place them back on her desk.
While the rest of us were worrying over getting our printing to
match exactly the example shown on the blackboard, Esther got
praise for any kind of mark that resembled the letters. Toward the
end of the second-grade school year, she had gone into some kind
of a seizure and thrown up all over the boy in front of her and then
fallen out of her chair and started to choke. Mr. Likens, the principal, came rushing in and pulled her tongue out and then put a fold
of notebook paper between her teeth. While the rest of us kids
cowered against the wall, Doc arrived and had Esther carted out.
She didn't come back after that. Somehow, it got to the Bykovskis
that I had been the one she threw up on, and my mother said Mrs.
Bykovski stopped her at the Big Store and said how sorry she was.
But I wasn't the one.

"What is on your mind, Sonny?" Mr. Bykovski asked, pulling my
thoughts of Esther away.

I took a deep breath and told him about going down in the mine
and then what Dad had said about Wernher von Braun and Mr.
Bykovski being a Jewish person. "Dad said I should ask you about
Dr. von Braun working for the . . . Germans and all." I couldn't
bring myself to say the word *Nazis*. I sensed it would be the same
as if I cursed in front of him.

Mr. Bykovski put his cup down carefully in the saucer, the
sound a minute, clinking noise in an otherwise silent room. "This is
a hard thing," he said slowly, as if he were having difficulty getting

his mind turned to my question. He drummed his thick fingers on the armrest. "Your Dr. von Braun," he said, still speaking slowly and carefully, "helped monsters, and for that he should be blamed." He set his mouth in a hard line. "There are concepts of forgiveness and redemption. . . ." He knitted his brow and shook his head. "For this, we need the rabbi, but he is in Bluefield and we are here." He sipped his tea and thought some more. "Sonny, please listen and remember you are hearing from a profoundly ignorant man."

Ike Bykovski spoke then of the way a man can change and how it is possible to forgive if not perhaps to forget. "This is not your sin, Sonny," he said. "It is Dr. von Braun's. If you're asking my permission to admire him for what he has become, you don't need it."

Mrs. Bykovski spoke from the kitchen. "Maybe there's a certain father who's jealous of a certain rocket man."

"Mary!" Mr. Bykovski admonished.

"You think my dad is jealous of Wernher von Braun?" I asked the empty kitchen doorway.

"I am sure Mrs. Bykovski is just suggesting the possibility of it," Mr. Bykovski said. He frowned at the kitchen and then came back to me. "And how are your rockets doing these days?" he asked, clearly wanting to change the subject.

I was ready to change it myself. "We got one up to nearly eight hundred feet. Next time, we'll bust a thousand, I know it!"

"That is very good! And your machine-shop work? Have you been practicing?"

"A little. But I think we need some more lessons." I explained that Quentin and I believed we had a solution to the erosion in our nozzles, but it would require machine work beyond our capabilities.

"I will speak to Leon Ferro," he said. "He could turn out such work in short order."

"Would you really talk to him? I wouldn't want to get you in any more trouble."

Mr. Bykovski shrugged my concern away. "Leon will want to trade. Will you be prepared?"

"Better than I was the last time," I said, remembering when O'Dell and I had gone looking for roof tin.

"While you're trading, we could use a new commode," the voice said from the kitchen.

"Mary!"

"Well, we could."

"I'll see what I can do, ma'am," I called out while Mr. Bykovski chuckled.

THE next week, Quentin caught the school bus to Coalwood and we went to the big machine shop. Mr. Ferro waved us into his office. "Yeah, Ike told me you were coming," he said, rearing back and putting his boots up on his desk. "Let's hear what you're after."

What we were after, I said, was some kind of steel for our nozzles that could withstand heat, pressure, and oxidation. "The steel we've been using burns up."

"Sounds like it also needs to be thicker," Mr. Ferro said. He had taken a pencil and, for no apparent reason, was balancing it on his upper lip. He rocked his head, working to keep the pencil from slipping off.

"Yessir," I said, mesmerized by his trick. "We think at least an inch thick. We need a hole drilled through its center too."

"SAE 1020 bar stock ought to do you fine," Mr. Ferro said, taking the pencil and tapping it against his temple before sticking it behind his ear. He looked up at the ceiling. "It has a high melting point and good tensile strength too. Expensive stuff though. Take some time to drill and shape. C'mon."

Quentin and I followed him through the shop, his machinists busy at their drill presses and milling machines and lathes. When we caught their eye, they stopped long enough to grin and wave at us. "Rocket boys," they mouthed to one another over the drone of their machines. Mr. Ferro stopped at a workbench and picked up a tap and threading tool. "I'd recommend inserting machine screws around the diameter of your thing to hold it in place. What did you call it?"

"A nozzle."

"We need a mechanism for sealing off the upper aperture too," Quentin said.

Mr. Ferro looked at me. "We need a top plug," I translated.

He nodded and took the pencil from behind his ear and pulled out a sheet of paper from the bench. Some of his men wandered over, peering over our shoulders. They were all grinning. "We gonna get into rocket-building, boss?"

Mr. Ferro handed me the pencil. "Draw me what you need."

I drew parallel lines to represent the casement and then showed the plug at the top and the nozzle at the bottom with a hole—about a third of the diameter—drilled through it. Mr. Ferro perused my effort. "Sonny, if you want work done in this shop, you're going to have to give me an engineering drawing. I'll need not only this side view, but also a top view and a detail on the plug and the nozzle. Think you can do that if I give you an example to follow?"

"Yessir, I can," I said. Except for winning Dorothy, which still remained my unsolvable puzzle, I figured I could do pretty much anything I wanted to do if only I worked at it hard enough.

We went back to his office. Mr. Ferro sat down behind his desk while Quentin and I stood. He eyed me. "Sonny, you know where I live?"

He knew I did. He lived in the group of joined brick houses below the Dantzler house known as the Apartments. I had delivered the morning paper to him for years, once managing to fling a folded *Bluefield Daily Telegraph* right through a tier of milk bottles on his front porch.

"Every time it rains, it turns into a mudhole behind my place," he said, leaning back and lacing his fingers behind his head. "Could sure use some gravel back there."

Machining and materials for gravel. Gravel, like all things in Coalwood, could be supplied by my father. After I completed my engineering drawing of the nozzle, there was nothing to do but to go up to the mine. Dad looked up from his desk when I entered his office. "I heard you've been talking to Ike Bykovski," he said. "And now you're visiting Leon Ferro. You get around, don't you?"

I was always astonished how he knew everything I did at almost the moment I did it. "Dad," I said, "I really need your help."

"You want gravel." He shook his head. "Leon Ferro's been after me about that for weeks. It's not going to happen. Get it out of your head."

"What can I do to make it happen?"

"Nothing. What's that you got?"

I showed him my drawing of the nozzle, the casement, and the top plug. He studied it. "That's not bad work," he allowed. "But you need to show the thickness of the tube." He showed me how to place the arrows and where to put the dimension mark.

"Thank you, sir," I said.

"Get out of here. I've got work to do."

I rolled up the drawing. "The gravel?"

He stared at me. "You don't give up, do you?"

"Mom says it's the Lavender in me."

Dad's left eyebrow shot up. "By God, I'd say it's the Hickam!"

I sensed an opportunity. "Dad, I'm sorry about what happened that day you took me in the mine."

"Mining's in your blood, little man," he shrugged. "I guess you'll figure that out, sooner or later."

"I still want to work for Dr. von Braun."

He nodded. "We'll see."

"The gravel?"

He sighed. "We'll see."

"And sir? . . ."

"What?"

"Mrs. Bykovski needs a new commode."

"Get out!"

Mr. Duncan installed Mrs. Bykovski's new fixture the next day about the same time the first of three two-and-a-half-ton truck-loads of gravel showed up in Mr. Ferro's backyard. I knew better than to thank Dad for any of it. Some dogs you're better off to just let lie in the sun. When I went to the machine shop, I expected to see the machinists hard at work on my rockets, but I was disappointed. Mr. Ferro explained he'd come up short on the steel tubing. "The tipple shop will have it," he advised without apology.

"But you promised *you'd* get it!" I complained.

"I sure could use some lumber for my front porch," he said, nonplused. "Got some rot out there."

Willy Brightwell was the name of the man who had taken Mr. Bykovski's spot in the tipple machine shop. I knew him fairly well. His son Willy, Jr., often played touch football with us, coming down with the other boys from Mudhole to scrimmage with us on the broad concrete between the church and the Club House. Mr. Brightwell shook his head at my request for steel tubing. "Naw, Sonny, I can't do that. Your dad, well . . . you know your dad."

I caught Dad at home with my latest request as he sat down and tried to give the paper a quick read. "No way," he said, rattling the paper and then jumping up at the black phone's insistent ring, "and that's final."

The tubing showed up on the back porch two days later, leaning against a far corner, along with some bar stock as well. I took it, no questions asked, still letting that old dog soak up the sun. If Dad wanted to pretend he wasn't really helping me, who was I to argue with him?

At the machine shop, Mr. Ferro presented us the finished product of our newest design, *Auk XIV*. Quentin hefted it while the machinists who built it circled around. "I fear the ratio of the mass of propellant added compared to the mass of the empty rocket will be too small," he said. "I have deduced that there is a relationship between these two masses that must be within certain parameters."

"He says it's too heavy," I told the machinists. I took the rocket from Quentin. It *was* heavy, and there wasn't much room for the propellant after the nozzle and top plug were bolted in place. The fins and nose cone would add more to the weight. I doubted even rocket candy could get this dense little rocket off the ground.

"What needs to be done is to increase the volume of the cylinder with only a small amount of additional mass," Quentin stated.

"It needs to be longer," I translated again.

A machinist—Clinton Caton was his name—raised his hand. "I'll do it, boss," he said.

Mr. Ferro nodded agreement. "It's all yours, Clinton."

Mr. Caton, as it turned out, was a man of vision. Without any advice from me, he lengthened the rocket to two and a half feet, a monster. To fill it took a pot and a half of rocket candy. While the candy was still soft, I pushed a glass rod into it—borrowed from Miss Riley's lab supplies—forming a spindle hole.

The following weekend, our rocket rocked in the stiff, frigid wind that swept over Cape Coalwood, enough so I was afraid it might be blown over. Sherman and Billy dragged out a six-foot steel rod O'Dell had found discarded behind the machine shop and jammed it into the slack beside the pad. We used a wrap of wire to make a loop at the top and bottom of the rocket and then slid it down the rod. Mr. Ferro's machinists crowded around, braving the bitter wind. Jake and Mr. Dubonnet were there too. "Might work," Jake said of the guide rod. "The rockets on my wing in Korea were on a short track that got them off straight."

"I heard you went to see Ike Bykovski," Mr. Dubonnet said, "and then Leon Ferro and then your dad for supplies. You're making the rounds, aren't you?"

I just shrugged. He already knew everything, anyway.

"Some of the boys in the union hall were wondering what John L. Lewis would think about UMWA members building rockets."

I didn't like the sound of that. Mr. Dubonnet could stop the machinists from working for me if he wanted to. I never knew what undercurrents between union and management might be coursing through Coalwood. "What do you think he'd say, sir?" I asked nervously.

He laughed his rich laugh. "I can just see those big, thick eyebrows dancing. He'd love it! Maybe I'll tell him. He might be interested in forming the United Mine and Rocket Workers of America!"

After we got Jake, Mr. Dubonnet, and the machinists safe in the blockhouse or hiding behind their cars, *Auk XIV* erupted from the pad, spinning once around the rod before hurtling into the sky. Quentin threw himself out of the blockhouse with his theodolite and started tracking. Sherman limped outside, scribbling notes on the flight. It angled slightly over in the direction of what we called

Rocket Mountain and kept climbing. It was our best rocket yet. When it was just a dot against the blue sky, it stopped and came hurtling down, disappearing behind the highest ridge on Rocket Mountain. We took off, scrambling up through the woods. Billy was in the lead. He was not only a good runner, but had a great nose for burnt rocket candy. A whole hour later, tired, our knees bloody from battering them against the rocks we had to climb, we found *Auk XIV*. It had landed nose-first full bore against the only rock outcropping within a hundred yards. Its casement was bent and its nose cone turned to sawdust. At least the nozzle was intact. There was erosion and pitting within, but it had held up. Quentin finally gasped up beside us; even Sherman could move faster through the woods than he could. He paused, his hands on his knees, trying to get his breath. Then he thumbed through Jake's trigonometry book. "Three thousand feet," he concluded.

Three thousand feet!

"I think we'd better give those boys a call down at Cape Canaveral," Roy Lee said. "We could teach them a thing or two."

The machinists waved me into their shop a week later to show off a rocket that they had built on their own. It followed the last design, except it was six inches longer—a three-footer. They had also put in the top plug and the nozzle with machine screws rather than welding them. Eyebolts were attached on the top and bottom for the guide rod. I accepted it gratefully, and the boys and I dragged out the hot plate and filled it up. The following weekend, *Auk XV* drilled into the sky to the applause of the watching machinists. I could tell it wasn't going to go as high as the smaller *Auk XIV*. In fact, it managed only half the altitude. Although the machinists were still thrilled by it, Quentin and I worried over its performance all week, trying to figure out what had cut into its altitude.

"We may have reached the maximum performance with rocket candy," Quentin said. "Perhaps there's a break-even point for all propellants."

"We need more tests," I said, "to be certain."

Quentin's face lit up. At last I had agreed with him. "My boy, although I have had my doubts, there are moments, such as this one, when I believe you are quite capable of learning. How about the science fair this year?"

"We're not ready," I said. "We still need a book so we'll know what we're talking about."

Quentin shrugged. "If we keep going the way we are, we can write our own book."

THE girls in the Big Creek band outnumbered the boys four to one. Although before the suspension the football boys might have had the hearts of the girls, we boys in the band had more ready access to them. There were no football games in the fall of 1958, but the band still took trips around the district on holidays to march through the various towns that invited us. It took two buses to carry the eighty band members and our equipment. Band buses tended to be real cozy, especially after we had performed and were coming back to school at night. Tired but happy, we sat in the dark bus, some of the more lucky boys paired off with the girls of their choice, smooching in the back. Dorothy was one of the saxophone players, and she always chose to sit beside me and even occasionally rest her head on my shoulder while I sat stock still, fearful of moving a muscle that might disturb her angelic rest.

The band members liked to sing quietly on those dark nights, the bus warm with radiating bodies. A favorite song was "Tell Me Why."

> *"Tell me why the stars do shine.*
> *Tell me why the ivy climbs.*
> *Tell me why the ocean's blue.*
> *And I will tell you just why I love you."*

I remember Dorothy's head shifting on my shoulder; she was mumbling something.

> *"Because God made the stars to shine.*
> *Because God made the ivy climb.*

Because God made the ocean blue.
Because God made you, that's why I love you."

What was it she said? Something to me she could say only in her sleep? I hoped so, was willing to pretend so. "I love you too," I said so low even I couldn't hear it, but my heart still thumped wildly at the audacity of it. The bus rolled on, filled with dreams.

DURING one of our study sessions in late November, I worked up the courage to ask Dorothy to the Christmas formal. "I wish I could," she said, shaking her head sadly. "That boy I went out with last summer asked me and I told him yes."

That boy, I knew, was a college student from Welch. Dorothy had told me all about him. "But you said he was mean to you!" I protested. "How could you agree to go out with him again?"

"Well, he asked me before I realized what kind of boy he was," she explained.

"And you're still going to the formal with him?"

"I told him I would, and I can't go back on my promise," she sighed. "But I'll be thinking of you, Sonny. I will."

The way she looked at me so pitifully, I couldn't help but feel sorry for her. The night of the big dance, I suffered at home, barely paying attention to the book—Steinbeck's *Sweet Thursday*—I had borrowed from Dad's bookcase in the upstairs hall. I stayed awake until two o'clock in the morning, relaxing only when I was certain Dorothy must be home and safe. I saw Roy Lee the next day. He'd been to the formal, and I couldn't resist asking him if he'd seen Dorothy. "I saw her," he said noncommittally.

"Was she . . . did she seem to be having a good time?"

He looked past me, off into the distance. "What do you want me to say?"

"Well, the truth."

Roy Lee put his hand on my shoulder. "She was all over that guy."

CHRISTMAS 1958 wasn't a white one, but it was bitterly cold, no problem in a town sitting on a billion tons of the finest bituminous coal in the world. As always, Mom bought the biggest tree in town, and Jim and I wrestled it indoors for her. Unwilling to remove an inch off it, Mom made us wedge it against the ceiling at an angle. When Dad came home, he silently got his stepladder out and cut two feet off the top. Mom hated his handiwork, saying he'd made it look more like a bush than a tree. After we decorated, Daisy Mae and Lucifer immediately began to pull down every bulb and bauble within paw's reach, and Chipper took up residence deep within, squawking at anyone who walked past.

On Christmas morning, Mom came into my room and sat on my bed, handing over a large manila envelope. Not having any idea what it might contain, I opened it and, to my amazement, found an autographed photograph of Dr. Wernher von Braun with a personal note in his own handwriting. I couldn't believe what I was reading. He was congratulating me on the success with my rockets, suggesting that I continue my education and perhaps I might one day find a job in the space business! The note ended by saying: *If you work hard enough, you will do anything you want.*

I stared at the photo and then the note, going from one to the other. I couldn't believe I was touching these things that the great man had also touched. "Mom? How . . ."

She grinned, a little proud of herself, I think. "I wrote him about you, Sonny. I thought he'd like to know who was getting ready to come help him build his rockets."

I grabbed her around her neck and gave her a big hug, surprising her as much as me. I had never had such a wonderful present! For the remainder of the Christmas break, I read and reread von Braun's note. I even offered it to Jim to read, but he claimed he didn't know who von Braun was. I tried to get Dad to read it too, and he said he would, but he never got to it.

I carried the von Braun photo and letter to school the first day back after Christmas. At lunch in the auditorium, Quentin fingered them like they were holy artifacts. "Prodigious," he whispered in reverence and awe.

13

THE ROCKET BOOK

ONE NIGHT IN January, it started to snow, a little at first, and
then steadily. Before I crawled into bed, I heard the muffled footsteps of the hoot-owl miners trudging through the deepening
snow. I looked outside and could barely see them through the
heavy snowfall. Daisy Mae huddled in beside me, purring. I
reached down and petted her and then fell asleep.

I woke to the sound of tire chains on packed snow. I saw
through my bedroom window that everything was white—the yard,
the road, the filling station, and the mountains. Only the tipple and
man-hoist kept their black visage, steam rising from the deep shafts
beneath them. I pulled on my jeans, shirt, and sweater and hurried
downstairs to the kitchen, where Mom already had the radio tuned
to WELC. Johnny Villani, the announcer, was cheerily commenting
on the snow, advising everyone to be careful, and, no, there was no
announcement of any school closing. Jim rose from the table as I
got there, grunted that it would be nice to take a day off to go
sledding, and disappeared upstairs into the bathroom. I gulped
down hot chocolate and toast, ran back upstairs to throw my
homework inside my notebook, then back down to balance books
and notebook on the banister post, and then to the television to
hear a few minutes of the *Today Show* with Dave Garroway. There
was little news on the space race, so when I heard Jim finish in the
bathroom, I took the steps two at a time upstairs, brushed my
teeth, and then rushed back down to grab a heavy coat from the

closet in the foyer. Jim was gone already, was in fact already climbing on the bus when I went out the front door. Mom chased after me, her housecoat pulled tight against the cold, and caught me just in time to hand me my brown-bag lunch. "Late again, younger Hickam!" Jack announced, giving me the eye. Then he saw my mother. "Mornin,' Elsie, how do?"

"I'd do better if I could get Sonny moving in the morning, Jack." She smiled up at him.

"Aw, the boy'll get some sense someday," Jack said, swinging the door shut. Mom waved and carefully scooted her house slippers up the walk.

I made my way up the aisle, wedging myself three to a seat beside Jane Todd and Guylinda Cox, already dozing. Carol Todd, Jane's cousin, and Claudia Allison wedged me in on the other side. As we trundled through town, I saw a few women out in their front yards shoveling coal into shuttles to carry inside to their Warm Morning heaters. Most of the women were bare-legged, and peeking beneath the bottom of their old woolen coats were pastel-colored nighties, standard Christmas gifts from miners to their wives during good times. Mom liked to tell about the time when she and Dad lived in one of those houses—just after they were married—and she ran out into the snow to the coal box with nothing on but her Christmas-night nightie and encountered a line of miners on their way to work. Naturally, they all stopped to comment.

"Now, Elsie, Homer will be buyin' you a coat soon, darlin'," Mr. O'Leary said sympathetically.

"He bain well better," Mr. Larsen added, outraged but eyes popping.

"Ah, that Homer," Mr. Salvadore said, putting his fingers to his lips, "he'sa lucky, lucky boy."

Mom grabbed her shuttle and ran for the porch, only to slip, both feet flying over her head, her pink ruffled matching house slippers sent sailing. At least the snow cushioned her landing, which was solidly on her backside. The miners started to climb over the fence to help her, but she told them to stop, dared them to take a step farther. She said she was fine, but she didn't make a move because if she got up, they'd see a lot more of her than she wanted any man to see, even my dad. So there she sat, melting the ice beneath her until the miners left—only after asking her many more times than she felt was necessary if she was sure she was all right—and then she made another run for the door. She was so embarrassed she didn't venture outside for the rest of the day, and when Dad got home after work, he found the Warm Morning cold.

"Why didn't you keep this fire burning?" he demanded, raising up the heater door and peering at the cold ashes on the grate. "I work hard all day, and I expect to come home and see something burning in here."

"You want to see something burning?"

"I sure do."

"Okay." Mom went upstairs and came down with the Christmas nightie and the matching slippers and stuffed them all into the Warm Morning and set them ablaze. "Better?" she asked. When Dad was within listening range of the story, he added to it by saying the house stunk for days afterward. Mom said a furnace was installed in our house the next year—the first for a house on our row.

A dozen more kids got on board at New Camp and then at Substation, Roy Lee among them. He had a speech assignment and began practicing to a thoroughly bored Linda Bukovich. Carlotta Smith got on at Number Six and all us boys perked up, watching her move sideways up the aisle in a short open jacket and tight sweater. It wasn't that she was exactly a beauty: She had a puffy, childish face marred by acne and her hair was stringy, but just the sight and scent of her could set a boy's heart a-pounding. Roy Lee leaned over and wiggled his eyebrows and whispered in my ear, "Old Glory," and, against my will, I laughed. It was shorthand for the cruel phrase for ugly girls with great bods: "Put a flag over her face and *you know what* for Old Glory." Not that we had a chance in the universe of doing any such thing. Carlotta could find no place to sit, so she stood beside me, her rounded bottom within inches of my face. Either guilt or embarrassment forced me to my feet, and she mumbled her thanks and pushed in beside Jane and Guy Linda, who woke just long enough to marginally scoot over for her.

"Woo-woo." Roy Lee grinned. He stood up long enough to whisper again. "You think being polite will get you some of that? What would your darling Dorothy think if she knew what you were thinking?" He sat down, sniggering. I tried to grab him, and he dodged and then broke out laughing.

Jack watched all our nonsense in his rearview mirror. "Younger Hickam," he barked, "you want to get thrown off again? No? Then get down here, right now!" I pushed my way through the dozing aisle to sit, at Jack's point, on the steps. Jack went through his gears, selected one, and we were off, crunching up the short straightaway before the first curve of Coalwood Mountain.

The fifth curve up was especially precarious. It canted toward a one-hundred-foot-high precipice without even a tree to slow the bus down if it went over. Jack braked. "Everybody out," he ordered. "Walk around that curve and go about halfway up the straightaway and wait for me. Leave your stuff."

He opened the door, and I climbed out with a busload of half-asleep students stumbling behind me. We trudged silently around the curve and kept going as ordered and then turned to wait for Jack. He eased the bus around the curve and then, gears changing, ground his way up to us. He opened the door and we all climbed back on. I took up station once more on the steps. Cresting the top of Coalwood Mountain, we were faced with a steep, straight stretch followed by a series of curves that dipped and turned. Jack slipped into a low gear and we trundled slowly through them, coming out at a short straight stretch that bottomed out into a wide, inside curve, a rocky cliff looming over it. I gazed with wonder at thirty-foot-long icicles hanging from the cliffs like crystalline stalactites.

We rolled with ease down Little Daytona and through Caretta, past the minehead there, and then up on War Mountain, where again Jack ordered us off the bus to walk around a particularly treacherous curve. We arrived an hour late for school. Mr. Turner was waiting for us at the door. "Go to your scheduled classes," he said. "Get the homework for the classes you've missed from your friends. Move, people!"

Before chemistry class started, Miss Riley called me to her desk. "I have something for you, Sonny," she said. "See me before you go home." With the excitement of watching the snow fall all day and school dismissed an hour early, I was on the bus home before I remembered that I was supposed to see her.

It kept snowing through the night. Lucifer moved inside, taking possession of the rug at the bottom of the basement steps. Dandy and Poteet also stayed in the basement, except for quick runs to the yard to do their business. The next morning, I crawled out of bed to silence. Nothing was moving outside except walking miners. Johnny Villani made the announcement: The schools were open, but the buses weren't running. If students could walk to school, they were expected to go. The rest of us had the day off.

I went into the living room for a rare treat, to watch the *Today Show* all the way through. But I got barely a glimpse of J. Fred Muggs before a snowball hit the living-room window. When I looked out, I saw O'Dell, Roy Lee, and Sherman with sleds. Jim and a knot of his pals had already grabbed their sleds and headed to Coalwood central to test the road between the church and the Club House. "Come on!" O'Dell yelped, so excited he was bouncing up and down. "We're going to Big Creek! Nobody's ever done that before on sleds. We're going to be the first ones ever!"

Mom was sipping coffee in front of her tropical beach. The palm tree was done, and it looked as if she were adding coconuts to it. "We're going to sled all the way to Big Creek," I said.

"Well, don't freeze to death," she sighed over the rim of her cup.

I trotted down the basement steps, carefully stepping over Lucifer, who raised a single irritated eye. Dandy and Poteet ran in circles, excited by my excitement. I found my sled and pitched it outside. Then I went back to my room and put on another pair of jeans, an undershirt, and a thick flannel shirt over it, two pairs of socks, my galoshes, and a heavy wool coat. No hat. It was not the style of teenage West Virginia boys to wear a hat, except the black-felt kind with a feather in it, and then only at dances. Mom saw me go and called me back, handing me a knit watch cap. "If you don't wear this, your brains are going to freeze," she said, and then waved at the other boys. "Are y'all crazy?"

"Yes, ma'am!" they chorused happily. "Come with us!"

"Not in this lifetime," Mom replied. I took the cap from her, put it on to satisfy her, and then whipped it off as soon as she shut the door. I stuffed it in my coat pocket. A few cars were managing to move, their chains clanking. We crossed to the filling station and waited until one headed toward Coalwood Mountain. Roy Lee grabbed its bumper, falling onto his sled, and one by one we formed a chain, hanging on the feet in front of us. I was the last sled. When the car pulled into the company store at Six, we let go. We trudged up the road and then set ourselves against Coalwood Mountain.

The snow we stepped through on the road was pristine, our tracks the first. There was good traction to it, and we were soon at the top of the mountain. We threw ourselves down on our sleds and, yodeling our delight all the way, flew down the dipping curves, slicing new double-runner tracks. We slid down Little Daytona and into Caretta. There, at the church, we chained onto another car and went all the way to the Spaghetti House. Others had walked up War Mountain, and we followed their tracks. We crunched past the little houses perched precariously on the nearly vertical slope on both sides of us. Then we slid down into War. We arrived at Big Creek High at lunchtime, leaned our sleds against the wall just inside the main door, and walked in as if we were kings of the earth. Mr. Turner caught sight of us. "If you boys think you're going to class, you're very much mistaken. The county superintendent has suspended classes for everyone for the remainder of the day. However," he continued, "please go to your teachers and get your homework before you leave."

I headed for Miss Riley's room and was relieved to find her at her desk. "I'm sorry I forgot to see you yesterday," I said.

I must have been a sorry sight, because she looked at me with sincere concern. "How did you get here?" When I told her, she held out her hand. "Let me feel your hands," she said. "Oh, they're ice cold. You go down to the cafeteria and get some hot chocolate."

I did as I was told. When I got back, she opened her desk drawer and withdrew a book. It looked to be a textbook. Its cover was red. "This came in yesterday," she said. "Miss Bryson and I put our heads together and ordered it for you. Here."

Miss Bryson was the librarian. I took the book and read its title, written in gold gilt on a black bar imprinted on the front cover. It was the most wondrous book title I'd ever seen:

PRINCIPLES OF GUIDED MISSILE DESIGN

I flipped through the pages of the book, seeing chapter titles, amazing chapter titles, passing before my eyes: "Aerodynamics Relating to Missile Design," "Wind Tunnels and Ballistic Ranges," "Momentum Theory Applied to Propulsion," and "Flow Through Nozzles." Then I read the most wonderful title of a chapter in any book I had ever held: "Fundamentals of Rocket Engines."

"There's calculus in there and differential equations," Miss Riley said. "You could ask Mr. Hartsfield. He might help you."

I reverently turned the book in my hands. "Can I keep it for a while?"

"It's yours, Sonny. You can keep it forever."

I felt as if she had just given me something straight from God. "I don't know how to thank you!" I blurted.

"All I've done is give you a book," she said. "You have to have the courage to learn what's inside it. Come on. You can walk me out to my car."

She put on her coat and I escorted Miss Riley down the hall, past Mr. Turner, who watched us suspiciously, and outside to the teachers' parking lot. At her car, she put her hand on my arm. "Sonny, it may take awhile, but I believe you can learn the things that are in this book. Then," she smiled, "maybe Quentin and I will finally convince you to enter the science fair."

"Miss Riley," I said, "if you want me to enter, I'll do it."

"When you're ready," she said.

At that moment, I believed I was ready for anything just because she believed I was. Coming back, the other boys ran up, carrying their sleds. "Come on, Sonny! We're going over to Emily Sue's to play hearts." I must have looked hesitant, because O'Dell added,

"Dorothy will be there too!" Roy Lee eyed me, looking unhappy. For some reason, he'd developed a dislike for the love of my life.

Emily Sue lived in a house built on the side of a nearly vertical mountain across the creek and not more than a hundred yards from Big Creek High School. Her father owned a big scrap yard in War, and her mother was the third-grade schoolteacher at War Elementary.

On this strange sort of school-but-not-school day, and with the snowy vista outside, Emily Sue's kitchen seemed twice as welcome, and warm, and fun. Her mother greeted us and then left us to ourselves. We sat around the kitchen table, drinking hot apple cider, eating homemade cookies fresh out of the oven, and playing hearts.

As O'Dell had advertised, Dorothy was indeed there, sitting across from me. I saw, almost as if for the first time, how gorgeous a girl she was. She had this great laugh, kind of a backward hiccup, that I found absolutely charming. Roy Lee nudged me. I followed him into the living room. "Will you stop staring at Dorothy like some kind of heartsick puppy? You're going to give me diabetes."

"What are you talking about?"

"She doesn't love you, you sap!"

I felt like slugging him. Instead, I said, "I bet I can get her to kiss me."

"When?"

"Now."

"This," Roy Lee said, "I gotta see."

We went back to the game. "Dorothy," I said, my heart racing, "I bet Roy Lee you'd kiss me. Today. Now."

Dorothy looked up from her cards, her mouth open. "What did you bet?"

"Just that you'd do it."

Silence fell around the table. Dorothy glanced at Roy Lee, who rolled his eyes. She put down her cards, stood up, and kissed me on the forehead. "There," she said.

"No good," Roy Lee said. "Got to be on the lips."

That wasn't part of our bet, but I didn't care to disagree with him. I looked at Dorothy expectantly. "I guess he's right," I said.

She took a short, unhappy breath. "Stand up," she said. I did, and she walked around the table and pecked me on the mouth. "There. Happy now?" She stalked out of the room.

"Dorothy?" I called after her.

Emily Sue snickered. "Fastest kiss in the history of the universe."

"See what you've done?" I snapped at Roy Lee.

He shrugged. "Me? The question is, do you understand what just happened?"

"Go to hell."

The hearts game broke up, and Roy Lee, Sherman, and O'Dell put on their coats for the journey back. "Come on, Sonny," Sherman said. "It'll be dark if we don't get going."

I looked at the closed door of the bathroom where Dorothy had disappeared. "I'll be along. Go ahead."

As soon as the others had left, she came out. I started to apologize. "It's that Roy Lee," she said, biting her lip. "He's such a greasy rat."

I showed her the book Miss Riley had given me. She made me sit down beside her on the couch so that she could look at it closer. "I'd like to learn calculus too," she said. "I want to learn everything I can."

The telephone rang, and Emily Sue said it was Dorothy's mother. She was on her way to pick her up. I followed her outside. My sled was leaning alone on the fence at the street level. Emily Sue's mother called from the front door. "I saw the other boys catch a ride in a truck, Sonny."

Dorothy's mother offered me a ride back through War. I climbed out of the car in front of Dorothy's house, retrieving my sled from the trunk. Dorothy got out with me. "Will you be all right?" she asked. The snow had started falling again.

"I'll have fun the whole way," I told her.

She looked around, as if to see if anyone were watching. Her mother had already parked the car and gone inside. Without warning, she hugged me and kissed me on the mouth, this time a lingering caress. "Be careful," she said, her sweet lips brushing my ear. "I don't know what I would do without you."

I stood in a state of rapture after she left. Two cars passed by, but I was too dazed to stick out my thumb. After that, no one came along and I started to walk. It was getting dark. Halfway up War Mountain, the wind started to blow and the snow pelted down so thick I couldn't see the lights of the houses in the valley below. My ears were freezing, and I remembered the knit cap. I pulled it out of my coat pocket and put it on, pulling it over my ears. Miss Riley's book was safe, comfortably pressing against my stomach, my belt cinched tight against it. I trudged on, leaning against the wind, until I reached the top of the mountain, and then I gratefully threw my sled down and slid to the Spaghetti House.

I walked on through Caretta. Any of the families there would have taken me in, but I wanted to go home. By the time I made it halfway up Little Daytona, I was thinking that I had made a mistake. The wind howled down the straightaway with near-hurricane force, almost knocking me down. My face stung from the driven sleet, and my eyelashes were coated with ice crystals. I considered turning around, but decided that I could make it if I just kept plodding along. I wasn't afraid, not yet.

It was terribly dark. I pushed on, finally picking up my sled and carrying it under my arm since it wasn't sliding very well in the deep snow. Somewhere near the top of Coalwood Mountain, I misjudged where I was and stepped off the road, disappearing under the snow into a deep ditch. When I finally clawed my way out, my pants and my coat were thoroughly wet. I could feel my pants freezing to my legs, and my coat felt as if it weighed a million pounds. For the first time, I felt fear. I was a long way from panic, but I knew what frostbite was—Coach Gainer had covered it in his health class—and I knew the danger of being wet in such cold temperatures. I peered down the road, hoping to hear the sound of a car or a truck, but there was nothing but silence. I had no choice but to keep walking.

When I reached the top of the mountain, I threw down my sled, but instead of sliding, it just lay there with me on top of it. The snow was too deep and sticky for the sled runners to slide. Groaning, I picked up the sled and walked into a near whiteout, feeling my way with each step. There were many places on the steep side of the road where there were no markers or fences. If I wandered off, I stood a very good chance of going over a cliff, and nobody was likely to find me until the next thaw. I stayed in the middle of the road. My teeth were chattering. I had to keep going.

I was shivering almost uncontrollably when I tripped and sprawled on my face. I lay for a moment with the thought that maybe if I just rested for a while, I would find the strength to go on. But I forced myself to stand. Coach Gainer had told us in health class how Arctic explorers had just gone to sleep when they froze to death. He claimed it was an easy way to die, but I didn't want to find out. I had rockets to build and Dorothy to win. Besides, even though I would be dead, I would never live down the story of me freezing to death on Coalwood Mountain. People would gossip for ages about how stupid I was. I got to my feet and shuffled on until I saw a single bare light bulb on the porch of a house built about a hundred feet down the side of the mountain. It was an old, dilapidated shack with a tar-paper roof. I knew some-body lived in it—there was usually smoke coming out of its stove-pipe—but I had no idea who it might be. I kept going. Southern West Virginia society did not allow for barging in on strangers in the middle of the night, no matter what the situation.

"Hey, boy, what you doin' out on a night like this?"

I peered through the swirling snow and beheld a woman carry-ing a lantern above her head. She wore a long cloth coat and galoshes. "I'm going home," I said, my frozen lips slurring the words. I couldn't feel my face, and my feet felt more like blocks of ice than part of me.

"Where's home?"

"Coalwood."

"You better come inside 'n warm up or you ain't gonna make it."

When I hesitated, she came after me, grabbing my coat. "Come on, boy!"

I gave in to her urging and followed her down a steep path to the little house. She pushed open a homemade wooden door and led me inside. An ancient pot-bellied stove glowed in the center of the room. A patched couch sat in front of the stove. A small rude

table was set under a window that looked down into the valley.
"Well, come on in!" the woman said when I hesitated. She doffed
her coat, kicked off her galoshes, and put on a pair of moccasins.
She took a pot off the stove and poured something into a cup. She
brought it to me. I saw her in the pale light of the lantern on the
table. She wore canvas pants and a plaid work shirt. She was about
thirty years old, I guessed, and had long straight blond hair. Her
thin face was plain but friendly. "Here's some sassafras tea," she
said.

I took the cup and drank it greedily, savoring the sensation as
the hot liquid flooded into my stomach. She took the cup from me
before I was finished. "We got to get you outa those wet things.
Everything off, let's go."

I hesitated, timid about taking off my clothes in the presence of
a stranger. "Oh, come on," she said. "You ain't gonna show me
nothin' I ain't already seen too many times."

She had a curtain—it looked like an old sheet that had been
sewn—slid across a pole nailed in a corner of the room for privacy.
She pointed toward it and I went behind it, pulling off my coat and
then my layers of shirts. I was relieved to see my book was still dry.
I laid it on a little two-drawer bureau in the corner and then handed
out my things to her, one by one. "I'll hang 'em up near the
stove," she said. She came back and swept the curtain back.
"You forgot your pants." I covered my chest with crossed arms.
"Gawdalmighty, boy. I ain't gonna do nothin' to you. Get them
pants off right now!"

I fought down my embarrassment and sat on a rough chair and
pulled my galoshes off and then my pants. "There, that wasn't so
bad, was it?" she chuckled, taking them. "I didn't bite you once!
You can keep your underdrawers. Gawdalmighty, them pants is
wet, ain't they? You ain't frostbit, are you?"

"My toes hurt," I confessed.

"Well, get them socks off too!" She added them to the line and
then came back and made me sit on the couch while she knelt in
front of me and inspected my feet. "Naw, you ain't frostbit," she
said. "Near though." She rummaged in a trunk and brought out a
long flannel shirt. It was a man's shirt, much too large for her. I

wondered how it was she had it. "Put this on and then sit there. Here's more tea. It'll help warm you from the inside out 'n that's the best way. Whose boy are you, anyway?"

I gratefully sat before the glowing stove, soaking up the delicious warmth. I wiggled my toes. They still hurt, but it was the pain of getting back to normal. "I'm Sonny Hickam. Homer and Elsie Hickam's second son."

"You're Homer Hickam's son?"

She said it in such a disbelieving tone of voice that I got worried. Dad had more than a few enemies. Was this woman one of them, or maybe her husband, or a brother? "Yes, ma'am," I said, and added carefully, "And Elsie Hickam's."

She drew up a chair and straddled it. "I know your daddy." She studied me. "I don't see it."

I was still in my underwear, and the way she was looking at me made me uncomfortable. I hunched over, trying to stretch the shirt over my lap. "Ma'am?"

"I don't see him in you. What's he doin' these days? How is he?"

That was the first time anyone had ever asked me to tell them about my father. I thought everyone just knew. "He's . . . okay. He works a lot. Mom got him an electric razor for Christmas."

"Did she, by God!"

"Yes, ma'am."

"Is he happy?"

Was my dad happy? Happy or sad were states I never thought applied to him. "I guess so."

My answer, vague as it was, seemed to satisfy her. "That's good. That's real good. My name's Geneva Eggers." She held out her hand and I shook it. It felt bony but warm. "Pleased to meetcha. I've known your daddy more'n a while. Say, you want some toast?"

Before I could answer, she got out a big black frying pan and poured some bacon grease in it out of a coffee can that was sitting on top the stove. She set the pan on the stove, over to the side, and then opened a bread box on a small table. She went back behind another sheet curtain and came out with two eggs. She cracked the eggs in a pot, whipped them with a fork, dipped in four slices of white bread, and laid them out in the frying pan on the stove. Just like that, the room was filled with the wonderful smell of hot bacon grease and eggs.

"Your daddy and me come from the same part of Gary Holler," she said as she cooked. "I followed your daddy around when I was in diapers. He took care of everybody up and down that old holler, always worryin' if the old people had enough coal for the stove or had food in from the store. He weren't no richer than nobody else, but your daddy was always wantin' to help." She looked at me. "You didn't know he knew me, did you?"

"No, ma'am." It was true. I had never heard of her.

"You know his daddy got his legs cut off in the mines, don'tcha?"

"Yes, ma'am. That's my Poppy."

"Well, that ain't all that happened to your Poppy. He got hit in the head in the Gary mine. He did! Big piece of slate caught him, knocked him silly. It was pret' near a year Mister Hickam didn't work. Your daddy just took over supportin' his family," she said. "Nobody never liked for nothin' when your daddy was around, I can tell you that!"

Geneva found a plate and stacked the French toast on it and set it on the table. She set out a pot of honey too. "Come on or it'll get cold."

The meal was delicious, somehow better for being cooked on the old stove. After I was finished I asked to use the bathroom. She gave me the lantern. "Put on your galoshes. Path is out the back door."

I knew all about outhouses. Mom's parents had retired to a farm in Abb's Valley, Virginia, and they had a privy. I followed Geneva's own footprints down the snowy path and found the outhouse at the end of it. It was a simple one-holer, complemented by the inevitable Sears, Roebuck catalog. It was too cold to dawdle. I did my business quickly and hurried back to the warmth of the shack. I found my clothes laid out on her narrow bed. "Pret' near dry," Geneva said, smoothing my pants. She went over to the stove and turned her back. "You can dress. Go on, now. You don't need to get behind the curtain. I ain't gonna look!"

I put the lantern down on the floor and then put on my layers of shirts and my pants, now warm and almost completely dry. I got my rocket book and tucked it under my belt. I looked up and saw that Geneva had been watching me. I wasn't sure for how long. "I-I really appreciate this, Mrs. Eggers," I stammered. Except for my mother, I'd never had a woman watch me get dressed before, and even that had been a while.

"It's Geneva, honey. Will you tell your daddy you saw me? Will you tell him I made you get yourself dry and fed you some toast?"

It seemed to me a most plaintive request. "Yes, ma'am. I will."

She helped me with my coat. "Tell him when your mom ain't listenin'. I wouldn't want her to get the wrong idea."

I didn't know what she meant by that, but I didn't question her. It wouldn't be polite. She led me up to the road, her lantern held high. Far below I could hear a truck trying to climb the mountain, its chains slapping the snow. If it made it, I knew I could sled down the packed snow left by its path.

"Tell your daddy, now. Will you promise?"

I nodded. "Yes, ma'am. I promise. Thank you for saving me."

"Hell, child, don't go on about that."

A big dump truck soon appeared, loaded with coal. I waved at Geneva and then fell on my sled in the tire tracks and slid quickly down the mountain, coasting all the way to the mine. When I got off the sled, I could see my house. There was a light in every window. As soon as I put a foot on the back porch, the door opened. I could see the worry in Mom's face, but she wasn't about to let me know it. "Don't track snow in the house," she warned. She looked me over. "You don't look any worse for wear."

Dad appeared, the evening paper in his hand. "I was about to get in the truck and go looking for you, young man."

I was filled with glory. "I went all the way to Big Creek and back on my sled!" I reached inside my coat and pulled out my book. "And look what Miss Riley gave me." I handed it to Dad.

He read the title and tentatively turned some pages. "It seems to be thorough," he said. The black telephone rang, and he answered it after handing the book back to me. "Get Number Two going if Number Three goes down!" he shouted, and I knew he was worrying about the ventilation fans losing power during the snowfall.

I went upstairs. Jim was in his room. I opened his door. He was lying on his bed, reading a magazine. "We sledded all the way to Big Creek," I told him. "Nobody's ever done that before."

"You morons went to school?" he growled. "We were supposed to stay home. Next time they'll make us all walk over there."

I went to my room. I snapped on the lamp on my desk and began to eagerly read my book, relishing the titles on each of the chapters, until I remembered Geneva Eggers. I went downstairs and found Dad slumped in his easy chair, back to his paper. Mom was in the kitchen. "Dad, a Mrs. Eggers up on Coalwood Mountain invited me inside her house and got me warmed up. She wanted me to tell you that."

Dad peered at me over the paper. "A Mrs. who?"

"Eggers. Geneva Eggers."

He studied me and then carefully laid the paper down on the footstool. "You were in Geneva Eggers's house?"

"Up on Coalwood Mountain. That shack off the road, about a third of the way down. She fixed me French toast. She wanted me to be sure to tell you about it."

The black telephone rang, but he didn't jump up to answer it. It was the first time I had ever known that to happen. He fixed his eyes on me. "What else did she do?"

"Nothing. Just got my pants dry."

"You took your pants off?" His voice was strained.

"She gave me a long shirt to wear."

He frowned. "And nothing else happened? You're sure?"

"Yes, sir. I'm sure."

Mom finally came in from the kitchen to answer the phone. When she answered it, there was a pause, and then she said, "I don't know, Clyde. He must have died or you'd be talking to him instead of me."

Dad kept looking at me as if trying to figure out if I was fooling with him in some way. Then he went to the phone and started shouting orders again.

It was still too icy for the bus to run the next day, but we rocket boys didn't try to cross the mountains again. We'd done it once and that was enough. We were sure to go down in the record book of teenage Coalwood heroes by the stunt. That night, Dad came into my room and closed the door behind him. "Let me tell you a story," he said, and I stopped reading my rocket book. He sat down on my bed. Whatever it was he was going to say, he didn't look happy about it.

"When I was a little younger than you," he said, "a house on our row in Gary caught fire. Those old Gary houses were just clapboard and tar paper. One spark and they burned up like straw. I was out in our backyard for something and saw the fire. There didn't seem to be anybody else around, so I went inside, thinking maybe there was somebody trapped in there. I tried to see, but it was too smoky. Then I heard this baby crying. I didn't know where I was, couldn't see anything. I just went by sound. I found this baby crying in all that smoke, like there wasn't anything in the world that was going to kill her! I picked her up and jumped out the window before the fire got us both. It turned out the whole family was in that house and I didn't see any of them, just the baby. Eight brothers, the mother and father, all burned to death."

Dad shifted on the bed, both his hands pushing down on the mattress. "I was kind of unhappy with myself for a long time after that. How was it I didn't see all those people in that little house?"

I just stared at Dad. I couldn't imagine why in the world he was telling me this awful thing, but whatever the reason was, I wanted him to stop. For a reason I couldn't define, I had a fear of knowing too much about him.

Dad looked me in the eye. "Anyway, that baby was Geneva Eggers."

"Oh" was all I could say. I thought of Geneva as a helpless baby, my dad carrying her to safety, and tears welled up in my eyes. I forced them back.

Dad flicked imaginary lint off my bed and looked up at the ceiling. He cleared his throat. "Sonny," he said finally, "what do you know of life?"

I didn't know what he meant. "Not much, I guess."

"I'm talking about . . . girls."

"Oh."

"You haven't ever . . ."

I flushed. "Oh, no, sir."

Dad focused on one of my model airplanes on my dresser. "I wouldn't tell anybody else about being in Geneva Eggers's house. She has sort of a business going there. Some of the bachelor miners—she's sort of their girlfriend."

I didn't understand. "Which one?"

Dad winced. "More than one . . . a lot of them. The occasional married man too."

My eyes widened, and I'm sure my mouth dropped open. I understood now. "She runs moonshine too," Dad said, his eyes still locked on the airplane. "Her husband got killed in the Gary mine five years ago. The police over there chased her this way. I gave her that old shack and I told Tag to leave her alone. Let her do what she wants." He got up and went to the door. "So now you know. Don't go see her again, and never, never tell your mother what I just told you."

Dad closed the door quietly behind him and I sat alone in my room, thinking about what he had said. I thought about the young woman in the shack, how she'd treated me so nice, and then I imagined what it must have been like for Dad to go into a burning house. I doubted I would have had his courage. I felt suddenly proud of him, more than for just his long-ago act of heroism, but because of what he had once been back in Gary and all that he had become because of his hard work.

The next day the school bus ran. I looked for Geneva Eggers often after that as we went by, going and coming. Sometimes she would be there, standing alone alongside the road. She studied the windows as the bus passed, smiling if she saw me. She didn't wave, nor did I. She was Dad's secret and I was hers.

14

THE PILLAR EXPLOSION

Auks XVI–XIX

NOW THAT I had received a letter from the great man himself, I
felt almost as if I were already on Dr. von Braun's team. On
February 1, I heard on the radio that the Russians had launched
Luna I, the first man-made object ever to break away from Earth's
gravity. The velocity required for that was 25,500 miles per hour, or
approximately seven miles per second, a distance I could visualize
easily because seven miles was the distance between Coalwood and
Welch. As the Russian spacecraft streaked toward the moon, I
went up on the Club House roof to use Jake's telescope just to see
what I could see. Jake didn't join me, because he was going out on
a date with Mr. Van Dyke's latest secretary, a redhead from Ohio.
Although Mrs. Van Dyke had threatened to hire an ugly secretary
for her Mister, somehow another lovely from up north had appeared. Jake called up to me. "See any Russians on the moon,
Sonny?"

I poked my head over the edge of the roof and waved. "Not yet,
Jake. See any down there?"

He threw back his head and howled at the moon just as the
redhead appeared on the Club House porch, her spike heels clicking. Jake took her in his arms, twirled her around while cupping
one of her breasts in his hand, giving me the grinning eye while he
did it, and then guided her into his 'Vette. He and the redhead
sped, with tires screeching, toward an unspecified destination. I
envied him and wished with all my heart that someday I would

learn how to have Jake's confidence and devilish enjoyment of life. Deep down in my heart, and with considerable sadness, I suspected I never would. Born and bred West Virginians, it seemed to me, had a suspicion of anything that was too much fun, as if it were maybe a sin.

I went back to the telescope and put my eye to the eyepiece. There was some speculation that the Russians' payload was a bucket of red paint. I spent the night on the roof alone wrapped in my heavy wool coat, alternately dozing against a chimney, and waking to look through the telescope. To my relief, no revolutionary red star appeared on the moon's yellow surface. The next day, the *Welch Daily News* said *Luna I* had missed the moon, but by a mere 3,728 miles. The next time they tried, the politicians and editorial writers worried, the Russians might hit it and then what kind of world would we live in? I worried with them. Would we never catch up to the Russians in space? Every time the United States launched a satellite, the Russians launched one bigger and better. I trusted that Wernher von Braun, at least, was doing something about it. In my own small way, I figured I was too.

I hurried every day after school to my desk to pore over my rocket book. On the weekend, Quentin hitched a ride across the mountain to study the book for himself. All morning he sat on the side porch and carefully, reverently, turned each page, his face a frown of concentration. I tried to sit with him, but there was so much I wanted explained that I knew I would be too much of a distraction. Chipper sat on Quentin's shoulder, his little black eyes following every turn of the page. His interest worried me more than a little. Chipper had, after all, eaten the family Bible the winter before, chewed right through it from Genesis to Revelation, shredding generations of inscribed Hickams in the process. Mom thought it was the cutest thing. All I knew was if I found tooth marks on one page of my rocket book, I was going to declare hunting season on a certain bushy-tailed rodent.

Mom fixed up some fried baloney sandwiches for lunch and called Quentin and me into the kitchen. Quentin continued to leaf through the book at the kitchen table. Finally, he said, "There's a lot of theory in here that presumes the reader already knows subjects we haven't had, chief among them, I suspect, thermodynamics and calculus. Did you note the discussion of isentropic and adiabatic flows?"

I slid my chair beside him. I had skipped over the chapter he was on, titled *Elementary Gas Dynamics*. "I don't think anything in there will help us build a better rocket," I said. Actually, its pages of equations had discouraged me.

"Perhaps not," he replied coldly, looking down his nose at me as if surprised at my shallowness of purpose. "But isn't it something you want to understand? All of these equations lead to a discussion of what happens to a gas when it enters a flow passage." He looked at me again. "Sonny, a flow passage and a rocket nozzle are one and the same!"

I looked perhaps blankly, because Quentin sighed and turned the page and pointed at an illustration: two trapezoids on their sides, the small ends facing one another. The figure was marked "Characteristics of Flow Passage for Subsonic and Supersonic Flow Expansion and Compression." "There it is," he said triumphantly. "The answer to everything. Don't you see?"

I peered at the illustration. "See *what?*"

"*Look!* This is how a rocket nozzle works, why it's designed as it is. Did you even bother to read about De Laval nozzles?"

I had at least done that. A Swedish engineer, Carl Gustav De Laval, had shown that by adding a divergent passage to a converging nozzle (one that necked down to a narrow throat) the expansion of the fluid (or gas) coming out of the throat would be transformed into jet kinetic energy. In other words, the gas came out of the passage faster than it went in. I told Quentin my comprehension of it and he nodded. "Yes, yes. You understand. Good."

"Then we should build . . ."

Quentin's face took on a characteristic smugness. "Perfectly calculated De Laval nozzles. We do that, Sonny, my boy," he reared back in the kitchen chair and waved his baloney sandwich, "and we're going to fly rockets not thousands of feet but miles into the sky." He took a giant bite of his sandwich and chomped on it thoughtfully, lettuce hanging out a corner of his mouth.

"If we learn how to work these equations," I said.

Quentin nodded. "Yes. That will be the trick."

I WOKE to an earthquake in the night, my heart slamming against my chest. Dogs were barking up and down the valley, a battery of yips and yelps and howls. The telephone rang in Dad's room, and then I heard his feet hit the floor. He raced downstairs and I heard the hollow thumping of the basement steps as he went down them. I looked outside and saw him heading up the path to the mine, struggling into his coat as he ran. He stopped once, coughed, and then kept going.

The mine was lit up, big spotlights playing around the grounds. A gathering murmur, people in their yards talking across the fences, coalesced outside. Mom went downstairs, pulling her housecoat around her. Jim and I, putting coats on over our paja- mas, followed her out into the yard. Mrs. Sharitz shared the news over the fence. A bump in the mine, a big one. That meant a pillar—or more than one—had exploded. I remembered what Dad had told me that Sunday in the mine, the energy that was concen- trated in pillars by the tons of rocks resting on them. But he had also told me that they were carefully engineered to hold the weight, and something had to be done wrong to make them explode. I pulled my mother to the side and told her what I knew. She looked at me sourly. "Your father will take care of it," she said.

"But something's not right," I said. "This shouldn't happen."

She huffed, exasperated at having to talk about it. "Sonny, I've been around coal mines all my life. What's supposed to happen and what does happen in them are two different things. You think Poppy was supposed to get his legs cut off?"

"But Dad said if the calculations are done correctly—"

"Don't you think Wernher von Braun does his calculations right too?" she demanded. "I still see his rockets blow up."

Mom wrapped her housecoat tighter around herself and walked away from me. After a while, the dogs stopped their howling, subsiding to a series of whimpers, and everybody went back inside. The next morning and most of the day Dad still didn't come home, but we knew, because the fence spread the word, that no one was hurt and that there was only one exploding pillar, in the gob far from the face. Dad had plunged inside the mine with the rescue team—they proudly called themselves the Smoke Eaters—and drove straight to the site of the pulverized pillar just in case there was anybody hurt. I found that out when Mom complained to him about it. I was in my lab and could hear them in the kitchen. "It's not your job, Homer," she said from the stepladder in front of her painting.

"I *trained* those men, Elsie."

"Then let them do what you trained them to do. You should stay back like Mr. Van Dyke."

"You just don't understand," Dad replied.

"Homer," she sighed, "the one thing in this old world I do understand is *you*."

THE next Saturday, the day clear and cold with light winds, we set *Auk XVI* on the pad, sliding it down the launch rod. A small group of people waited expectantly on the road. Basil sat on the hood of his Edsel. I was also surprised to see some Sub-Debs, in their distinctive leather jackets, standing apart. I went over to them when I saw Valentine Carmina, drop-dead gorgeous in a tight black skirt and a white V-necked sweater.

"I just *had* to see your rocket, Sonny," she said mischievously, taking me by my arm and walking me away from the other girls. They were smoking and giving some boys who were yelling at them the finger. "I can't take 'em anywhere," she sighed, looking at her company.

"I'm glad you came, Valentine," I said, feeling suddenly very warm. Her breasts had pretty much swallowed my arm.

She released me and turned to study me. "Sonny, I got something to say and I'm going to say it. I know you're crazy about that Dorothy Plunk, but she don't seem to give a flip about you. Boy as cute as you shouldn't have to put up with that." She smiled and winked. "You should have a girl who 'preciates you. Now, I ain't sayin' who that girl oughta be, but you need to take a look around."

Roy Lee walked up to us while I was trying to figure out what to say. My tongue seemed to be tied in knots. "I hate to break this up—believe me, I do," he said, "but we've got a rocket to launch." He led me off, his hand firmly on my arm. "Now, *there's* a woman," he said.

I worked to clear my head. Then I wondered if Roy Lee had been talking to Valentine. Before I could make an accusation, I heard Quentin whoop. He was downrange with one of the old mine telephones O'Dell and Roy Lee had acquired from the mule barn and for which, I remembered, we still owed Mr. Van Dyke. We were testing them for the first time. Sherman and O'Dell had run phone wires all morning, hooking them up to old truck batteries donated by Emily Sue's father from his scrap yard. When I entered the blockhouse, the speaker to the phone squawked, startling me. "Blockhouse," Sherman said into the phone and then listened. "It works, it works!" he yelled.

We each took our turn at the phone, talking to Quentin. "You ready?" I asked, all excitement.

"Ready!"

"Stand by." I looked around the group. Roy Lee went outside and ran the BCMA flag up the pole. When he came back, I started to count, with Sherman on the phone to Quentin keeping up with me. "Ten—nine—eight—seven . . ."

At zero, I touched the bare tips of the ignition wires to the battery. There was a spark as the wires touched, and then *Auk XVI* suddenly jumped off the pad and flew straight up the rod into the sky, a white plume of rocket-candy fire and smoke sizzling behind. It had a fine contrail, allowing us to track it all the way. When *Auk XVI* was just a pinpoint in the sky, it arced over smoothly and dropped downrange. I heard a satisfactory *thunk* when it hit the slack.

We fired three more rockets that day, two two-footers and one three-footer, counting up to *Auk XIX*. All performed flawlessly, flying nice elliptical trajectories downrange to impact on the slack. Billy aimed his theodolite from a spot beside the blockhouse, and Quentin did the same with his downrange. Two observation points made the trig more accurate, and Quentin calculated that the two-footers reached an altitude of around three thousand feet, the three-footer around two thousand feet, an observation that confirmed our suspicions on rocket performance and their size. When going for altitude, bigger wasn't always better. Basil stood beside us while we talked, taking notes.

I heard the toot of a horn and saw the Sub-Debs leaving. Somebody was waving something pink out of the back window of their car. They were panties. "I wonder why it is that females wear such creations," Quentin mused while the rest of us boys stared, openmouthed. "They seem too slippery for good posture in the seated position."

"Shut up, Quentin," Roy Lee said.

"I have also wondered why they wear stockings as separate items. If they were combined with the, *ahem,* panties, it would be more efficient."

"Shut up, Quentin," we all said as one. Basil laughed, but wrote it all down.

When we returned from Cape Coalwood for a BCMA meeting, brother Jim contemplated us sourly from the sofa, where he was watching television. "Will you sisters hold it down?" he griped as we chattered about our rockets.

Quentin had a copy of the *McDowell County Banner* containing Basil's latest story about us. Jim snatched the paper from him, looked it over, and then threw it on the floor. "Why would anybody want to write something on you jerks anyway? So you shoot off rockets, so what?"

"People show jealousy in a lot of ways," Quentin snapped. "In your case, Jim, old chap, it's overt."

Jim turned to look at me. "You better have your jerk friend take that back or I'll slug him."

Quentin shoved his fists in the air and made little mixing moves. "Come on, big boy. Anytime!"

"I could smash you with one hand tied behind my back," my brother said.

Quentin barked out a laugh. "And I could outsmart you with one brain tied behind mine!"

Jim reddened and came off the sofa. He pushed me out of the

way, and he might have gotten to Quentin if Roy Lee hadn't stepped in front. Roy Lee was no match for my brother, but he gave me time to get up and wedge in beside him. Together we might do Jim a little damage, if only accidental. "Sister morons," he muttered and went back to the sofa.

"Let's get out of here while we still can," I whispered to Roy Lee, and he, Sherman, O'Dell, Billy, and I shepherded Quentin, still sputtering, upstairs to my room. Chipper ran in past us and leapt up on the window curtain and hung there. I passed the rocket book around, inviting all the boys to inspect the pages of equations. "To get all that we need to know from this book," I said, "we're going to have to learn calculus."

"And differential equations," Quentin added.

"Are you two crazy?" Roy Lee demanded. "We can hardly do the homework they give us now."

"Nevertheless," Quentin said, "it must be done."

"I want to learn calculus," Sherman said simply, and then O'Dell and Billy said they did too.

Roy Lee sighed. "Here we are, a bunch of Gawddamned West Virginia hillbillies wanting to be Albert Einsteins."

"Wernher von Brauns," I corrected him.

"Same thing," he said, but the way he said it I knew he was with us.

15

THE STATE TROOPERS

MR. HARTSFIELD PUSHED my book back across his desk. "You disturb my lunch with this nonsense? How can you expect to learn calculus when you didn't understand algebra?" That was meant for me. The other boys had made A's in algebra.

Quentin interceded. "Sir, we've been studying trigonometry already on our own." He produced Jake's book. "We needed it to find out how high our rockets flew. But it's unlikely we could teach ourselves calculus. We need your help for that."

Mr. Hartsfield looked sympathetically at Quentin. "Perhaps *you* could absorb the material," he said, but then he hung his old gray head. "But, no, I see no purpose in it."

"We need it to learn how to build better rockets, Mr. Hartsfield," I said. "It's for our futures, don't you see?"

Mr. Hartsfield softened for an instant, his watery eyes glistening, but then he snapped back to his usual dourness. "I've heard of your group, Mr. Hickam. Mr. Turner has spoken of it, and not in positive terms."

"What if we got Mr. Turner's permission for you to teach us?" I asked.

A smile almost played across Mr. Hartsfield's face. "I shall do what Mr. Turner tells me to do, of course. But surely you must know there is no hope for this class, no hope at all."

"Why not?" Billy demanded.

"Because," Mr. Hartsfield sighed, looking down and shaking his head, "this is Big Creek High School. Maybe if this was Welch High, the county superintendent would approve such a class, but not here. We're a football and a coal miner's school, and that's all we've ever been."

We were outraged. "That's not fair!"

Mr. Hartsfield looked up sharply. "Who ever told you boys life was fair?" he demanded.

"WELL, the pipe-bomb boy," Mr. Turner said from his desk. "And Miss Riley too? I hope you're not here to tell me you've blown up chemistry class."

Miss Riley told him our purpose and showed him my rocket book. "The boys are very serious about this, Mr. Turner," she ended.

"I presume Mr. Hartsfield has agreed?" he asked.

"If you approve it, he'll teach us," I said.

"Artfully put, Mr. Hickam," Mr. Turner observed, an eyebrow cocked. "Miss Riley, do you really believe this to be a good idea?"

"Yes, sir."

He tapped a finger on his polished desk top. "I see you have much to learn about school administration. I could not allow this course even if I wanted to. The county superintendent would first have to approve it, and I can assure you he won't. 'R. L., you are putting on airs!' he'd say." He waved at us with the back of his hand. "That's all. You're excused."

Miss Riley taught our class without her usual ebullience, the corners of her mouth turned down. My mom would have said that she "had her Irish up," and since she was Irish, I could sense that she was a little dangerous. On the way to English lit., I saw her emerging from the teachers' lounge, Mr. Hartsfield in tow. He was staring at the floor and shaking his head from side to side. She caught my eye and gave me a wink.

The next day, Quentin and I were called from typing class, told to report to Mr. Turner's office. Mrs. Turner, the principal's wife and also his secretary, was clearly flustered and leapt from her chair when we arrived. She ushered us inside. Turning from the principal's office window were two uniformed men, the patches on their arms identifying them as West Virginia State Police. "That's them," Mr. Turner said, and I knew we were in trouble deep.

The state troopers were huge and daunting in their gray uniforms. One of them advanced on us, holding a scorched metal tube with fins attached to it. "Recognize this, boys?" He held it out to us and we stared at it.

"It's yours, isn't it?" Mr. Turner accused.

Quentin recovered first. "May I inspect this device?" The trooper handed it to him, and he turned the tube over in his hands. "Interesting, is it not?" he asked me. "Note carefully how the fins are attached. See? They're spring-loaded. Ingenious design!"

"What is it?" I asked, finally finding my voice.

"Oh, come, boys," Mr. Turner said. "You must tell the truth. I think you know very well what this is. It is one of your so-called rockets."

"No, sir," I said. "It isn't. But these fins . . ." I took the thing from Quentin. I wasn't certain what I was looking at, but whatever it was, I wanted to compare the area of the fins with the area of the tubing. I suspected there was knowledge there to be gained. "Can we have this?"

The trooper snatched the tube back, his face clouded with anger. "No, you can't have this! It's evidence. Your rocket started a forest fire. It burned the top off Davy Mountain and almost got to the houses on Highway 52."

I remembered reading something about the fire in the *Welch Daily News*. There had been some speculation that it was caused by arson. "We didn't do it!" I yelped.

"I read in that grocery-store newspaper all about you boys," the other trooper said, ignoring my denial. He had a big square face and eyes looking for the lie. "You're the only kids in this county who are shooting off rockets, so it has to be you." He took out handcuffs. "Now, come along. We need to take you over to Welch to the courthouse. You are officially under arrest, both of you."

"There's a couple more boys involved," Mr. Turner said. "I'll have them called too."

Miss Riley suddenly appeared. "Why are you scaring these boys?" she demanded, inserting herself between me and the trooper holding the handcuffs.

"They tried to burn half the county down," he said.

"With this rocket," the other trooper added.

"Where was this fire?" she asked, in a doubting tone of voice.

"Davy Mountain. Between Coalwood and Welch by the way a crow flies. Or a rocket."

Mrs. Turner entered the office, holding a map of the county. She looked at her husband. He knew (and she knew he knew) that she had called Miss Riley to our rescue. "Maybe this might help?" she said to his scowl and then retreated. The one sure thing was there was going to be trouble in the Turner household later.

"Come over here and show us where your rocket range is," Miss Riley said to me. She spread the map on Mr. Turner's desk, his neat stacks of documents tumbling into heaps. He stood up, aghast at Miss Riley's forwardness, brushing imaginary debris from his vest.

I leaned over the desk, my trembling finger finding Coalwood and then moving down the valley toward the river at Big Branch. "Here," I said, finding the low place that was Cape Coalwood.

The troopers looked, and then one of them slapped Davy Mountain with his big paw. "You see, only an inch away!"

"An inch is ten miles on this map," Miss Riley said sardonically.

Quentin had been looking at the tube. "Of course!" he piped. "I should have known it the moment I saw these fins. They're spring-loaded because they have to snap out when this device leaves its storage tube."

Everyone in the principal's office turned to see what he was talking about.

"It's an aeronautical flare. I thought it looked familiar. I was reading a book on the civil air patrol just a month or so ago." Quentin perused the map. "Look here. There's the Welch Airport, just beside Davy Mountain. This must be a flare dropped from an airplane!"

The troopers looked at the tube, and then took it away from Quentin and looked at it some more, snapping the fins closed and back open again. Then they looked at one another and then at the map. Then they looked at us and then we all turned and looked at Mr. Turner, who seemed to shrink before us. Steadying himself, he carefully turned the map around and studied it through his half-glasses. He straightened up. "I think you'd better leave," he said quietly to the troopers. "Miss Riley. Would you please stay? And you two . . ." It felt as if his eyes were piercing me. "I believe you are supposed to be in typing class."

A week later, Mr. Turner called Quentin and me back to his office. This time, no troopers waited for us. Mr. Hartsfield looked up from a chair along the wall. "Big Creek will offer a class in calculus. The first class will begin in two weeks. It will be limited to six students. The superintendent said five, but I insisted on six so all the boys in your pipe-bomb club could be in it. A sign-up paper will begin circulating immediately." He stood up. "All right, you can go. You have gotten what you wanted. But heaven help you if you waste Mr. Hartsfield's time!"

A few days later, when I was next called to Mr. Turner's office, he had his hands folded on his desk and was looking grim. "I told you only six students would be allowed in the class," he said. "But seven signed up."

Mr. Turner tapped the stack of folders. "I'm afraid your grades were the lowest of the seven." He studied me. "Mr. Hickam, you have just learned a great lesson of life and this is it: Life is quite often ironic. You worked to get this class and now you will not be able to take it."

I stood there, my mind spinning, my stomach sinking. "Can you tell me who got the class instead of me?" I finally managed.

"Dorothy Plunk."

I wandered from Mr. Turner's office and into the hall, feeling weak, frustrated, and pitiful. I was tempted to run to Miss Riley and beg for her help in overcoming this terrible injustice. Why not seven students instead of six? But I didn't. Mr. Turner had kept his promise. Dorothy had the better grades and deserved the class. I remembered when I had shown her my rocket book and she had said how much she, too, wanted to learn calculus. She deserved her chance as much as I did.

"I'll teach you calculus," Quentin said.

We were going out onto the frozen football field behind Miss Riley and the tenth-grade biology class. Mr. Mams had asked her to perform a chemical experiment that had to do with the decomposition of organic materials. Quentin said it involved some chemicals he wanted to see. "I don't know, Q," I answered. "Why don't you learn it and work the equations? You don't need me."

"Nonsense," he spat.

Miss Riley sprinkled a little hill of gray powder in the grass. "This is zinc dust," she said. She poured sulfur on the powder and mixed it up with a stick. "You've smelled rotten eggs? That's sulfur dioxide, released from the chemical reaction of rotting organic material. This is going to stink just like that." She touched a long match to the pile, and it erupted into a huge boil of hot light and smoke.

"Yewww," the other kids groaned, holding their noses. They had had enough. They stamped around and shivered until Miss Riley led them back across the field.

Quentin and I remained behind. "Sonny," he said. "I think we've got our next rocket propellant."

I had been impressed by the massive amount of smoke and gas the combination had produced, but I was confused. "Q, why do we need a new propellant? What's wrong with rocket candy?"

"You are fortunate to have me as your scientist, old boy, or else I think you'd still be blowing up little aluminum tubes." Quentin could still be obnoxious when he wanted to be. "Haven't you been paying any attention at all? We've got the maximum results we're going to get from rocket candy. No matter what we do, we're not going to gain any more altitude. We need a new propellant."

I kicked at the debris. "What did you say this stuff was?"

"Zinc dust and sulfur."

"Hot stuff," I said.

Quentin cocked his head and nodded, as if he were a teacher who had finally gotten the correct response from the class dunce. "Indeed," he said.

As the winter of 1959 waned and the snow and ice melted, many miners who had been cut off the year before were called back to the mine. The Ohio steel mill had received a big order and needed coal, lots of it. For the first time in years, the company went to a three-shift, seven-days-a-week production schedule. New cars, heavy with fins and chrome, were parked in front of miners' houses from New Camp to Frog Level, and new swing sets, brightly painted, were erected in the backyards. Women and children wore new clothes. Living rooms glowed with new television sets, and home telephones—from the phone company, not the mine—appeared on little tables. Dad was in a frenzy every time I saw him, hopping up and down to yell into the black phone, running up to the mine in the middle of the night. Mom kept working on her painting in the kitchen. She was putting in a house on the beach.

About the same time, a strange traffic began at home: college football coaches, come to entice Jim to their schools. Despite his fears, the year's suspension had not dimmed his reputation. I was banished from the living room when they came, but listened all the same, sitting on the steps in the darkened foyer. "For heaven's sake, Homer, remain calm," I heard Mom tell Dad in the kitchen when the coaches from West Virginia University were there. "You're going to have a heart attack."

"Do you realize who's in our house, woman?" he demanded. He loaded up a tray with glasses of iced tea and rushed back to the living room, glancing my way as he passed. He had a huge grin on his face, but it disappeared at the sight of me. I must have looked particularly unhappy. "And what's with you?" he asked warily.

"My grades weren't good enough to get into calculus class," I shrugged. I told him the story quickly because I could see how eager he was to get back to the coaches.

He studied me. "Let me get this straight. You fought for a class, got it, and now you can't take it?"

"Too true," I replied.

"You want me to talk to Mr. Turner?"

I shook my head. "No, sir. He did the fair thing."

Dad nodded. "Yes, he did. I'm happy that you recognize that." He went into the living room with the tray, and soon I heard his laughter.

I trudged up the stairs. At the top of the steps in the hall there was a huge bookcase, six shelves tall. I turned on the hall light and found myself looking at the books, an idle activity. Then my eyes lit on one. It was titled *Advanced Mathematics, a Guide for Self-Study*. It was well-thumbed, some of the pages dog-eared. Its index included several chapters on differential equations and calculus. I found a piece of yellowed notebook paper within, calculations on it in Dad's handwriting. I realized I was holding the book he had used to teach himself the mathematics he needed for his job. Then I wondered why he hadn't mentioned it to me. Maybe, I fumed, he didn't think I could learn what was in it.

Reverend Lanier once preached that when a door is shut in our faces, we shouldn't worry, because someday, if we're properly patient, God just might open another. My mother, never the patient sort, had a different idea. If a door closes, she amended to me after his sermon, find yourself a window and climb through it. I took Dad's book—it was *mine* now—into my room.

16

A NATURAL ARROGANCE

Auk XX

DURING THE LAST week of March 1959, Dad went off to Cleveland, Ohio, for a mining-engineering conference. He was to give a presentation on ventilation, a great honor for a man without an engineering degree. It was a strange sensation knowing that he was not in Coalwood, not even in the same state. I felt uneasy without him nearby, but I wasn't sure why. In my nightly prayers, I always included by rote Mom and Dad, Jim, my uncles and aunts and grandparents (whether they were already in heaven or not), all the soldiers, sailors, and marines, Daisy Mae, Lucifer, Dandy, Poteet, and Chipper too. For the entire week, I added a special request that my father come back safely from his long journey.

My prayers worked, and Dad came home with a paper bag full of gifts. Mom got a faux-pearl necklace. Jim got a pair of binoculars. I got a fountain pen. The night after his return, Dad came upstairs and peeked into my room and asked me what I was doing. "Studying calculus," I replied. I didn't really want to discuss it with him, because I was sure he'd criticize me for wasting my time.

"You told me Mr. Turner wasn't going to let you take calculus," he said in an accusing voice.

"I'm learning on my own," I answered, and reluctantly showed him the book I was using.

He frowned. "That's funny. I don't remember you asking if you could use my book."

To divert him, I asked a question that I needed answered anyway. I pointed at an equation that defined the slope of a line. "I don't understand this little triangle," I said.

"You're not getting anywhere if you don't understand *that*," he said. "It's called a *delta*. Delta means change—the difference between one value and another over time." He went down on one knee and took the pencil from my hand. "You see, if the *y* coordinate and the *x* coordinate change, the point they describe changes too. Then if you change the time period—" He stopped in mid-sentence. "If you're not getting your class, why are you learning calculus on your own?"

"Dad, we're doing good down at Cape Coalwood. Come see us."

He stood up. "Well, maybe when I have time—"

"You always have time for Jim," I blurted, surprising myself as much as him with my vehemence. I let out a nervous breath. "Just come and see, Dad," I said. As much as I detested myself for it, I heard pleading in my voice.

He opened the door. "I still haven't given up on you being a mining engineer. We could work together."

I shook my head. "I don't want to do that."

"You don't want—little man, when you grow up, you're going to find out there's a lot of things you're going to have to do whether you like it or not."

"Yes, sir, I know—"

"But what I think doesn't mean a hill of beans to you, does it?"

"That's not true!" How could I explain? I struggled to find the words to tell Dad that just because I wanted to work for Dr. von Braun, it didn't mean I was against him. And why couldn't he be as proud of my wanting to build rockets as he was of Jim wanting to play football? Jim was leaving Coalwood too, wasn't he?

"You're not going to do anything I ask you to do, are you?" Dad accused.

"Dad, I . . ." I couldn't find the words. I cursed myself for my awkwardness in front of him.

He gave me a look of such disappointment that tears came to my eyes. Then he left, firmly closing the door behind him. A tear rolled down my cheek. I wiped it away with my shirt sleeve. It disgusted me. How could I let Dad get at me like that? I knew he didn't understand what I was doing, but I was right, wasn't I? The future was somewhere else, not in Coalwood, and I had to get ready for it. That's what Mom believed and a lot of other people too. But if I was so right and my father was so wrong, why did it make me feel so bad? If he'd just come to Cape Coalwood and see . . .

Despite my disgust, the tears kept coming. As always when I needed to clear my mind, I went over to my bedroom window and looked outside. Daisy Mae joined me, nuzzling my wet cheek. I could see miners moving up the old path, their lunch buckets glinting from the lights bolted high on the man-hoist. Other men were coming down the path, their work done. Every one of those men knew exactly who he was and what he was supposed to do. I wondered if the day would ever come when I would be able to say the same. I was sincerely beginning to doubt it.

I ARRIVED home from school the next day and found a note on my desk: *Mr. Ferro called—machine shop. How about we countersink the nozzle, one side or the other or both? (Sonny—do you know enough to answer?) Love, Mom.*

I sneaked a call in on the black phone to Mr. Ferro. I told him I thought countersinking the nozzle was a good idea. That meant the removal of cone-shaped material from each end, saving weight. I figured we'd get a little increase in altitude as a result.

"The men thought you'd like it, Sonny," Mr. Ferro said. "As a matter of fact, Caton already did it. Got you a three-footer ready, nozzle countersunk forty-five degrees both ends. You want to come get it? And also we were wonderin' if you might launch this weekend?"

When I said yes, Mr. Ferro yelled my answer to his machinists and I heard a whoop. "Tell 'at rocket boy we'll all be there!" someone yelled, and then I heard them do a simulated countdown. "Five—four—three—two—one—*whooosh!*"

I biked down to the big machine shop, finding the three-footer laid out on a black cloth on a table in the rear of the facility.

Besides the countersunk nozzle, Mr. Caton showed off his new design for attaching the fins to the casement. He had constructed a flange that ran an inch past the length of the fin on both ends and bent it to match the curve of the casement. Two narrow straps cut from cold-rolled sheet steel were lapped over the extended flange and used to clamp the fins solidly in place. As beautiful as the design was, I worried over its weight. I also worried whether Mr. Caton had made too many changes at once. "Naw, Sonny," Mr. Caton said. "It would take you forever to find the best design if you only made one change at a time."

I understood Mr. Caton, but I knew it would spur an argument with Quentin. Quentin preferred making our design changes one at a time so if we had a failure, we'd know its likely cause.

Mr. Caton used red paint to letter *Auk XX* down the length of the shiny steel casement on one side, *BCMA* on the other. The wooden nose cone was also painted a bright red. We stepped back and admired our creation. It looked like a first-rate, professional job. At school, I consulted with the other boys and it was agreed we would load *Auk XX* with melted rocket candy the following Friday and fire it on Saturday. Quentin caught the school bus to Coalwood to help me do the loading and then spend the night at my house so we could get an early start. When he saw all the changes Mr. Caton had made, his lip went out. "He may be a first-class machinist, but he knows nothing about scientific principles," he said. "To be successful, we must carefully test before making such modifications."

I told him about my conversation with the machinist. "I think he's right," I said. "It'll take us forever your way."

"And when this rocket blows up and you don't have a clue what caused it?" Quentin asked, his face pinched. "What will you have learned then?"

"The men down at Cape Canaveral say they learn more from a failure than a success," I countered.

"Then those men are full of crap."

"Is that what you think of me too?" I barked at him. "That I'm full of crap?"

"No, Sonny," Quentin replied calmly. "I think you're in a hurry, but for what I have no idea."

I knew how to cheer him up. "That's right. I am in a hurry. You should be too. We're going to enter the science fair next year." The truth was, I had been quietly giving it some thought all along, especially with Miss Riley often asking me if I'd decided to do it. She had been so good to me, getting me the book and all, that I wanted to please her. But what had truly pushed me into it was my anger toward Dad. If we won the science fair, that would show him, wouldn't it? I could wave whatever medals and ribbons we got under Dad's nose. If we lost, I would be no worse off than I was.

Quentin immediately brightened at the prospect. "Are you serious? That's wonderful. We'll win it all—county, state, and national! I know it!"

I put down the spoons I'd been using to measure the ingredients of the rocket candy. "Win it all? I thought you liked to take things one at a time."

Quentin stared at me. "Sonny, your parents can put you through college, can't they?"

I wondered how much I should tell Quentin about the kind of war being waged between my parents. "Mom's talked about it," I said cautiously. "I think she'll see that I go if I want to."

"Well, my parents can't pay for my college—they just barely keep me and my sisters in clothes and corn bread. I don't know about the other boys, but I bet their parents can't do much for them either. Yet I know—and they all know too—that somehow we're all going. You're the key, Sonny. You're our ticket to college."

"Me?" I felt like he'd handed me a big sackful of rocks. "Quentin, I don't know if they give out scholarships at the science fair, even at the national level. Medals and ribbons and such is all we'll get. It's for the honor of it, more than anything."

He shook his head at me, ever the patient teacher. "Do you not understand how our audacity will be perceived? Do you not understand, even now, what we have already accomplished, us coal miners' kids down in deepest West by God Virginia? Perhaps there may not be a direct scholarship prize, but success will get us noticed by someone. It's our chance, Sonny. *My* chance."

The imaginary sack of rocks got even heavier. I thought our rockets were to help us get jobs someday with Dr. von Braun, but Quentin was saying they were for something much more immediate. I started to tell him to forget it—it couldn't possibly work out—but then I thought of Dad. I had my own reason to win, didn't I?

When I turned back to loading rocket candy, I felt something I had never felt before—powerful, confident, and angry all at once. It seemed a natural arrogance that felt *good*. "All right, Q" was all I said.

AUK XX zipped up the launch pole, accelerating a hundred feet in an instant, climbing to two hundred feet, three hundred feet, hurtling vigorously toward the blue spring sky. Quentin and Billy jumped to their theodolites, but just as they got the angle on the speeding rocket, it exploded and steel shrapnel rained down on Cape Coalwood while we boys scrambled back inside the blockhouse and the machine-shop crew backed away on the road. When we heard the last *thunk* of rocket remnants hitting slack, we went out and began collecting the pieces. The machinists morosely gathered around the blockhouse. I carried the largest piece of the rocket, a center section of the casement, to Mr. Caton. Our proud lettering along the side was scorched, showing only the *k* of the *Auk* and only one *X* of the *XX*. "Was it the countersink that made it explode?" he asked.

"I don't think so," I replied. "The blowout seems about a third up the casement."

"There's no way to tell what happened," Quentin muttered, bringing in a handful of fragments. "You changed so many things at once. Who knows?"

Mr. Caton sorrowfully inspected the casement piece, running his fingers along the shredded ends. "I never thought there was anything powerful enough to blow out a steel tube like this. Bursting pressure's got to be in the neighborhood of twenty thousand pounds per square inch, even with the weld."

"What weld?" Quentin demanded.

Mr. Caton shrugged. "I used a steel tube with a butt weld down the length of it. We don't have much of the seamless kind. Too expensive. We used what we had on your last rockets." He turned the casement over. "There's the weld, right there."

I looked and knew the reason for the failure when I saw a deep rent right at the nearly invisible stripe of weld. A butt weld was where the two ends of the steel sheet that made the tube were simply pushed together and welded. Such a joint was too weak for our rockets. There was just too much sustained pressure.

The other machinists crowded in. "A lap weld would've been better, Clinton," one of them said. "Give you another ten thousand pounds of bursting pressure." A lap weld was where the two ends of the sheet were overlapped and then welded.

"Yeah, and I got some of that," Mr. Caton said sadly. "I don't know what I was thinking."

"Can we get seamless tubing?" I asked, displeased at working with any kind of weld.

"I'd have to order it," Mr. Caton replied doubtfully. "Your dad would have to approve."

"That's fine," I said firmly. "I'll take care of it. Get your order going."

I was certain that it was indeed okay for Mr. Caton to write up his order. If Dad didn't let me have the materials I needed, I'd still get them, one way or another, no matter what it took—guile, tricks, or outright theft. I didn't need Dad. I let every juicy morsel of anger and bitterness well up inside me, making no attempt to staunch any of it. Instead of hating the feeling, I gloried in it. I was becoming tough, just like *him*.

VALENTINE

IT WAS THE golden age of rock and roll, even for us kids in deepest West Virginia. At night, when we could hear stations far from our mountains, we usually tuned to a station in Gallatin, Tennessee, that played hearty black rock and roll. Although we didn't buy the Rezoid Royal Crown hair dressing that was hocked in between the songs, we came to love Chuck Berry, LaVern Baker, the Coasters, Fats Domino, Shirley and Lee, Ivory Joe Hunter, and Joe Turner. When Elvis and Carl Perkins and Jerry Lee Lewis came along, we listened to them too, usually on WLS out of Chicago, but our hearts were always with the black groups Gallatin brought us.

And also anything Ed Johnson played.

"Where's Ed playing this weekend?" was the question we antici-pated every Friday. Ed Johnson was the man who guided the classes of Big Creek High School through the golden era of rock and roll. A Marine who had island-hopped across the Pacific from Tarawa to Iwo Jima, Ed had seen the world by the time he was twenty and come back to West Virginia to forget it. He worked awhile in the mines and then hired on with the high school as a custodian. He married twice and had children, although we were never certain of the number. The day would come when he would leave West Virginia and move to Florida, where he would be elec-trocuted while cleaning a swimming pool. But while he was with us, Ed Johnson was a one-man recreation department in blue jeans and a V-necked sweater, playing for our dancing pleasure all the latest records on his homemade hi-fi system.

Ed's favorite place to play his records was the Dugout, which was actually the basement of the Owl's Nest restaurant just across the river from the high school. The Dugout was sparsely furnished—booth benches with backs on them along the walls, and support pilings scattered throughout the room that cut into the dancing space. It was also dimly lit, just a few pink and blue light bulbs in the low ceiling. There was a furnace with a coal pile beside it over in one corner. The Dugout had nothing going for it, except that we loved it. It was our little piece of rock-and-roll heaven. You could always tell how much you'd danced at the Dugout if you pulled off your socks that night and looked at the black ring around your ankles from the coal dirt.

Ed despised country music—to him it was "plinky-plunk and tacky"—and therefore so did we. He played a mixture of slow and fast songs, the fast ones played if they had a good beat. He rarely played Elvis, because Ed considered his music too fast to dance to and overly commercial. The slow songs he favored were heart-tugging: Dean Martin's "Return to Me," Billy Ward's rendition of "Stardust," "Chances Are," by Johnny Mathis, "It's All in the Game," by Tommy Edwards, and anything by the Platters. The last song Ed played at the end of his dances was always "Goodnight, My Love," by Jesse Belvin. Ed chose his music carefully, his dances having an opening act of excitement and greeting, a middle of pulsating, very danceable songs interspersed with romantic interludes, and then the inevitable ending. So powerful was the selection that couples clung to one another during "Goodnight, My Love" as if it were not only the dance that was ending but their lives.

Sherman called on a Saturday in April, saying we needed to take a break from rocket-building and go to the Dugout. I agreed. I was pretty worn out anyway, trying to be tough and arrogant like Dad. I just wanted to be a normal kid again and go to a dance and be with other kids too.

I told Sherman I thought maybe we could catch a ride over with Jim since he had the Buick, but by the time I got around to asking, he'd already taken off. Mom said he'd spent even longer than usual getting ready for his date, so she had to be somebody "pretty special." I didn't say it to Mom, but they were all "pretty special" to Jim. He had cut a wide swath through the girls of Big Creek High, leaving behind him a crowd of broken hearts.

Sherman and I stood in the dirt on the other side of the gas station for only a few minutes before someone picked us up and took us to Caretta. After a few more minutes with our thumbs out, somebody else came along and took us all the way to War. There, we kicked down the street, easing into the Sweet Shoppe and eating a hot dog and talking to the rosy-cheeked man behind the counter, who complimented us on our pink shirts, black draped and pegged pants, white socks, and brown loafers. We were a proud pair, dropping by the pool hall, playing a quick game of eight ball, and then, as the sky darkened, swaggering down the sidewalk. Susan Linkous, a pretty ninth-grader at War Junior, waved at us from her front porch and we "howdy'd" her, strutting even more.

A block away, we could hear the muffled music coming from the Dugout. The dance had been in full swing for an hour. At the door, we were greeted by a blast of warm air and the vision of shadowy, dancing bodies. Ed, who knew everybody by name, greeted us. His latest girlfriend, a blond young honey, stamped a black spot on the back of our hands after we gave her our quarters. Sherman saw a girl he liked and tapped her on the shoulder and hit the dance floor for about ten straight. His weak leg made him dance a little funny, one foot turned out for balance, but it didn't keep the girls from wanting a turn with him. I edged into the darkness and saw Emily Sue and Tootsie Rose sitting on a bench and went over and talked to them. Connie Peery was home from college, and she snared me for a quick dance while her boyfriend was sneaking a smoke in the parking lot across the river. Ed was into the opening sequence of his middle act, the records bouncy ones with good beats. I watched cheerleaders Cathie Patterson and Sandy Whitt dance with their boyfriends. Cathie was an energetic, athletic dancer. Her boyfriend's shirt was soaked in sweat from trying to keep up with her. She waved to me and I waved back. She called out, "How're your rockets doing?" and I nodded to say they were okay and so was I. I was having a good time, all my worries about rockets and everything else put aside. Rock and roll and being surrounded by my classmates and friends were good medicine.

Out of the corner of my eye, I saw Valentine with Buck Trant. I had been surprised during the last few weeks to see them sitting by themselves in the morning and at lunch too. That usually meant something serious was going on between a boy and a girl at Big Creek. I didn't think Buck was anywhere near good enough for her. I had no right to be jealous, of course, since I was committed heart and soul to Dorothy.

Valentine and Buck were apparently having a quarrel. She spun on her heel after shaking her finger in his face and stomped out of the Dugout. Buck hurried after her, whining, but Ed stopped him, said a few words, and I watched as the huge boy trudged sullenly back to a bench and slumped down. Sensitive to the ebb and flow of teenage romance, Ed often chose slow songs to get a couple back together. He was usually successful. I saw it more than once—the boy and girl, eyes closed, draped onto one another, swaying to Ed's romantic music, all forgiven. Ed made no such attempt for Buck. The boy was not one of his favorites. After a while, Valentine came back in and pulled Buck up for a fast dance. Buck shuffled after her, his big hands hanging limply at his side.

Connie's boyfriend was back, so Emily Sue and I did a fast dance, and then I danced with Becky Hurt and Tish Hampton and Mary Grigoraci and Dana Beavers. I asked Malvey Sue Harlow, a tenth-grader, for a slow dance, and then went back out on the floor for a "chicken" with Lucky Jo Addair—one foot out and then the other, our heads bobbing in time to the beat. Bodies swirled around me, and I inhaled the wonderful, intoxicating aroma of sweat and perfume all mixed up. Sherman had disappeared into the shadows with his sweetie, but I was sticking to the lighted area, making myself available. That was when I saw Dorothy.

I had never seen Dorothy at the Dugout before. She was standing alone just inside the door, wearing a black skirt and a pale green sweater with the collar of a white blouse peeking out. She had her dancing shoes on. I perked up, thinking that she had come alone, but then I saw her date enter behind her with his two quarters in his hand. I recognized him, recognized his flattop of blonded hair with the ornamental frontal curl, the snarly lips, the athletic lumber.

My brother took Dorothy's perfect little hand and together they walked out under the pink and blue lights and green and white crepe paper and began to dance to a booming Ed Johnson middle-act song, syncopated by the sound of my broken heart shattering on the concrete floor.

How I made my feet and legs move I do not know, but they got me to an empty bench in the back, where I sat as Jim worked his magic stuff on Dorothy in front of my eyes. I was a fascinated spectator in the way the surviving passengers must have been as they watched the *Titanic* sink. Other dancers swirled around the floor and some girls even came and asked me to dance, but I did not respond. I was too busy dying a thousand deaths. And then, oh, my God, Ed played a slow song, "It's All in the Game," by Tommy Edwards.

> *Many a tear has to fall,*
> *But it's all in the game.*
> *All in the wonderful*
>   *game,*
> *That we know as Love.*

Dorothy melted in Jim's arms. As my stomach tightened, he took his stubby hands and clasped them around her petite, perfect waist, while she tucked her head into his lumpish shoulder, her summer-sky-blue eyes closed, a contented smile on her perfect lips.

> *Once in a while, he won't call,*
> *But it's all in the game.*
> *Soon he'll be there at your*
>   *side,*
> *With a sweet bouquet.*

I saw Jim lean back, pretend to hand Dorothy a bouquet, and when she took the invisible flowers, I felt my soul curl up and die, and then all the blood that was in my body drained completely to my feet. I was numb and in exquisite pain at the same moment.

"Sonny?"

It was Valentine.

"Would you like to dance?"

I looked up at her and then at her hand held out to me. I took it instinctively and she walked backward, pulling me to my feet. She bumped into Jim and Dorothy. Dorothy opened her eyes sleepily and Jim scowled, but they moved out of her way. Valentine draped her arms around my neck. We swayed to the song, and then Valentine's lips were brushing my ear and I wasn't really thinking about Dorothy and Jim anymore.

> *Then he'll kiss your lips,*
> *And caress your waiting finger-*
> *tips.*
> *And your hearts will fly away.*

I cannot say that I remember leaving the Dugout with Valentine. When I thought about it later, I did seem to remember Ed patting me on the back in a special, urging way. Valentine and I had to cross the bridge to the parking lot in front of the high school, but I don't remember that either. I do remember Buck's old Dodge and Valentine opening the back door and getting inside, scooting across the bench seat in the back, and then reaching across it to take my hand and draw me inside with her. Then she locked all the doors and settled back, her hands moving to the bottom of her sweater, her arms crossing, and she raised it over her face. She shook out her hair as she dropped the sweater into the front seat. She smelled of musk and desire. Or was that me? She opened her arms and took me in.

I thought I heard Buck knocking on the window and wailing something, but not much more. Valentine had found WLS, and the DJ was in a romantic mood.

> *Love is a many-splendored thing.*
> *It's the April rose that only grows in the early spring . . .*

Santo and Johnny's "Sleepwalk" was playing when I next came up for air, the windows of the Dodge steamy and gray, the heavier droplets making curving little translucent streaks down them. I rested on her, my cheek so tight against her it was as if it were welded to her breast. After a while, she eased me up and out of the car. She leaned across the seat on her knees. "Sweet chile," she said as she hugged me. She touched me on my nose. "Other women will have you in your life, but nobody will have you first except me, and don't you ever forget that." And then she closed the door and I knew it was time for me to go. I walked unsteadily back to the bridge where Buck waited, his arms resting on the rail, his head hung over the water. I stopped beside his bulk as if I were in a dream. It was crazy, but I felt no threat from him, even though he had to know it was me in his car with Valentine.

Valentine started the Dodge and turned it around in the parking lot, then drove over the bridge and kept going without any apparent intention of coming back, even though it was Buck's car. Buck turned to watch her go. "Oh, how I love that girl," he moaned. All of a sudden I felt so sorry for Buck. He had been robbed of playing football his senior year, robbed of going to college, and probably robbed of ever making something more out of himself than what he was. I wanted to tell him how awful I felt, to console him somehow, to make it all better for him, even though it was me who had just been with the girl he loved. The best I managed was to pat his arm in a there-there way while he sobbed, his hands to his face. I stayed with him until finally it occurred to me that at the end of his crying, Buck might decide to throw me off the bridge. I fled into the darkness.

AT the Dugout, Ed had already played "Goodnight, My Love," and the place was empty. Sherman was gone, most likely catching a ride to English to the drive-in before hitching the rest of the way home. I looked at the clock inside the Owl's Nest and was surprised to see that it was well past midnight. There was a storm in the air, and as I hurried down the sidewalk, the first patter of rain came, and in the distance, I saw lightning. A car eased up the road and I stuck my thumb out and it stopped. It was 2:00 A.M. before another ride got me all the way to Coalwood. The rain was lashing the valley, and thunder and lightning crashed across the mountains.

The tipple area of the mine seemed to be strangely lit—a big spotlight directed at the man-hoist—as my ride carried me past it. When I looked down the valley, there was a light on every porch, and I could see the dark shapes of people walking up the street toward the mine. The back door of my house stood wide open. I came inside cautiously and found Mom sitting at the kitchen table in front of her painting. She stared at me and spoke to me as if she were the guard to the very gates of hell. "You are not to go to the mine," she intoned gravely as lightning split the air and turned her face bluish-white. "No matter what else you do this night, you mustn't do that."

18

THE BUMP

I COAXED OUT of Mom what had happened. Three hours before, two fans had been struck by lightning, and thirty minutes
after that, a bump had occurred near the face. Then of course the
black phone rang, and Dad heard there was a fall, men were hurt,
maybe trapped, methane was surely seeping in. If the fans didn't
get going soon, there was likely to be an explosion that would
streak through the length of the mine. Dad ordered everybody out
who could get out and then slammed the phone down and ran to
the basement.

"I told him not to go inside," she said bitterly. "Let the rescue
squad do it, I said. But no—it's not his way. 'I have to go,' he said.
I told him he just couldn't stand that somebody else might go
inside his precious mine and do something without him.

"So now you're home," she said to me. "I'm not even going to
ask you where you've been. Go to your room and go to bed.
That's where Jim is. What goes on inside that filthy pit doesn't
concern either one of you."

As ordered, I went to my room and looked out the window and
saw cars and trucks rushing past, making toward the mine. Then I
saw an ambulance coming across the mountain from Welch.
Thunder boomed and lightning flashed and the rain came in
swaths. Townspeople straggled by carrying umbrellas, their coats
pulled tight around their throats.

I couldn't stand to hide in my room and not know what was

happening. I opened my window and went out on the roof and then swung down over the ledge, caught the windowsill, and dropped into the yard. I hopped the fence and merged with the line of people winding up the path toward the mine. Nearly all of the people of Coalwood were gathering there. A barrier of saw-horses had been erected, and the women whose husbands were inside stood in a special place behind them. I heard people talking. The rescue team had descended hours before. There had been no word since.

I watched from the shadows of the bathhouse. My thoughts were in turmoil—so much had happened already this night. Doro-thy was gone forever. I could never feel about her the same way I had after she'd been with Jim. I watched the men and women of the Salvation Army share a prayer with anyone who seemed to need it. I thought of Valentine, and it gave me no pleasure. My first experience with a girl seemed sad. Valentine had loved me with pity in her heart.

Doc waited at the ambulance with its crew, and Mr. Van Dyke watched with a little knot of foremen and engineers from the porch outside Dad's office. Jake was there too. Little children stood around everywhere with their parents, as quiet and stoic as the grown-ups. A baby wailed behind the sawhorses, and a Salva-tion Army lady took it and rocked it quiet while its mother sagged in the arms of another woman.

The storm quieted, and there was a murmur of excitement when the man-hoist winch creaked and the lift came up, but it contained only a few men of the rock-dust crew. They reported the rescue team had gotten near the face, but a loader covered by the fall stood in their way and they were trying to pull it out. That gave the men in the crowd something to chew over. I ventured over to listen to a couple of wheezing retirees trying to explain everything to Doc. "They'll manhandle the rock off that loader, Doc, and then they'll tie on a cable, try to use a motor to pull it out."

"Why don't they blast their way through?" Doc wanted to know.

"Can't, might set the black damp off," the old miner said, using the colloquial term for methane. "Or it might cave in the rest of the roof. Naw, diggin' them out's all they can do now."

"How long will that take?"

"Couple hours, maybe more. Depends on how much slate they got down. Them ol' boys got a chance if they can get in there pretty quick. There's air aplenty the way this mine has the hell ventilated out of it. Story I got was the fans are back up. They just got to open up a little hole. Naw, Doc, they got a chance. You wait and see!"

The night dragged on, the rain ending and the clouds scudding away and the stars blinking on, one by one, looking cold and distant. A breeze rustled the budding trees above us on the mountainside, but, like everybody else, my focus was entirely on the silent tipple and the frozen winch wheel on the man-hoist. It seemed as if the shaft sighed every time steam rose from it, as if emoting a whisper of anticipation. Mr. Van Dyke came back out on the porch after talking on the telephone, and a rumor flashed through the crowd: One in the rescue team had been hurt, but they had broken through. Some men had been found dead. The wives behind the sawhorses dropped their heads and prayed quietly. The winch on the man-hoist creaked, and everyone looked up as it slowly began to turn. Doc and the Reverends Lanier and Richard walked toward the shaft and stood at the gate as the cable, rigid with tension, slid by. The people tensed, instinctively knowing that this was what they had been waiting for.

On the rising cage were two members of the rescue team, identified by the green crosses taped to their helmets. With them was a stretcher, the body aboard covered by a gray blanket. A miner opened the gate and held it back as they walked the stretcher off. Doc raised the blanket and took off his hat and said something to the wives. A path opened in their ranks and a woman, her arms wrapped around her old coat as if she were freezing, stepped through it. She walked regally behind the stretcher to the ambulance. When she crossed into the light from the bathhouse, I saw that it was Mary Bykovski. I couldn't stop the groan that escaped my lips. *Please, God, I want this nightmare to end.*

I started to go to her, but a voice stopped me. "No," Mom said. "Not now."

When she came out of the shadows, her eyes shot holes in me. I started to say something to her, probably some cowardly plea for understanding, but before I could she slapped me as hard as she could in the face. I rocked from it, my cheek burning and tears welling in my eyes from the surprise of it as much as the pain. Her face was twisted with anger. "I *told* you not to come up here," she said.

I stood my ground. "I was worried about Dad."

"No, you weren't," she hissed. "You don't worry about anybody but yourself. That's the way you've always been—*selfish!*" She turned in disgust and stalked away from me, going into the crowd and out of my sight.

I slumped against the bathhouse, my hand to my cheek as if it were glued there. My mother's opinion of me kept ricocheting in my brain: *selfish.* The ambulance carrying Mr. Bykovski started up and slowly rolled down the hill toward the road. I watched it and prayed: *Make it stop hurting. Please, God have mercy, make it stop hurting.* My prayer caught in my throat. Mr. Bykovski's body was in that ambulance, and my prayer had been for myself. Mom was right. I had always been selfish. There was yet another reason to be disgusted with myself.

The man-hoist winch creaked again, and the waiting wives shuddered as if a cold wind had blown through them. It seemed an eternity, but finally a dozen men came up, their faces black as the night sky. Some of them were being held up by others. A rescue-team member stepped out on the ground. He looked at the wives. "All alive," he said in a loud voice.

All alive! The wives pushed up against the sawhorses and then knocked them over to get at their men. Some fell and picked themselves up, but they kept going. They flew into their husbands, oblivious to the greasy coal smearing their clothes. The children crowded around, clinging to their fathers' legs.

Then I saw my father coming off the lift alone. His helmet was gone and there was a bloody bandage covering his right eye. He walked stiffly to Mr. Van Dyke. The general superintendent came

down off the porch and solemnly shook his hand. Then all the
rescue squad crowded around Dad, patting him on the back. He
dully accepted their accolades and then walked clumsily away, as if
his boots were made of lead. My mother left the crowd, but didn't
go to him. She just followed. I think she knew it was important to
him that he walk home under his own power. I waited until they
were down the hill to the road and then I followed, my cheek
where my mother had struck me still on fire.

Mom and Dad were in the basement—I could hear the shower
running—when I slipped inside the house and went to my room.
I heard them come up the steps and then the sound of bed-
springs giving as he was let down on his bed. Mom went downstairs.

Then the black phone rang. It seemed ten times louder than I
had ever heard it. Mom came running into the foyer, but instead
of picking it up, I heard her tear the phone from the wall, open
the front door, and throw it out into the yard. I came out of my
room, worried that she might hurt herself.

The black phone in Dad's room was still ringing. She came two
steps at a time upstairs, and pushed aside Jim—groggy with sleep
and wondering what all the commotion was about. She brushed
by me and threw open Dad's door, raised the window in his
room, and tore the remaining black phone from the wall and
threw it out too. "Go get Doc," she ordered me. I started for the
stairs, but Doc had already arrived, stalking into the foyer and up
the steps. He said nothing to me, but took Mom into his arms.
"It's going to be all right, Elsie," he said.

"Since when?" she choked. The two of them went into Dad's
bedroom and closed the door.

When Doc came out, Jim and I were waiting in the hall. We
had said nothing to each other. There was nothing between us to
say. "I had to put twelve stitches in his forehead," Doc reported.
"The cable they were using to pull out the loader snapped and hit
him in the head. The force of it cracked his helmet in two. He'll
probably lose his right eye. I'll have him over to the hospital
tomorrow and we'll know more then."

Doc walked to the banister and turned and looked at us. "A

dozen men would have died tonight if it hadn't been for your dad. That's something a son should know."

I followed Doc out to the back gate. "What about Mr. Bykovski?" I asked.

"He was operating the loader that got buried."

It was all too much. I couldn't take it anymore. I hung my head and started to cry. Doc put a steadying hand on my shoulder. "What's this?"

"It was my fault! Mr. Bykovski wouldn't have been there except for me!" I told Doc the story. "And if he hadn't been at the face, he would still be alive," I finished, my voice spastic with choked sobs. I looked up at Doc, meeting his eyes.

"Stop your sniveling," Doc hissed. "Gawdalmighty, don't you understand the kind of place this is? The men in this town go in that pit and hold hands with death every day."

I couldn't stop my tears, and they shamed me. They ran in rivulets down my cheeks, dripped off my chin. I hated myself for them.

"Ike built your rockets," Doc said resolutely, "because he wanted the best for you, the same as if you were his son. You and all the children in Coalwood belong to all the people. It's an unwritten law, but that's the way everybody feels."

He walked to his car and climbed inside, started the engine, and rolled the window down. "I'm going to tell you what your father would say if he could. Don't let me ever see you act like a sob sister again or, by God, I'll whip you myself. Coalwood is no place to be weak, but if you are, keep it to yourself and get the hell out of here as soon as you can."

I stood at the gate and watched Doc drive off. I looked up the road toward the tipple and saw the people coming back. I heard what sounded like normal conversation. Somebody even laughed. Mr. Bykovski was dead and my dad was maimed and they were acting as if it were all a relief. Only one man was dead. *Only one!* One man was killed in the mines in McDowell County all the time. The prayers at the shaft had worked well enough. I loathed them all as they passed me by, loathed that what passed for courage and endurance was in reality apathy even in the face of death. I wanted no part of them. I wanted only escape, to show my back to Coalwood forever.

Clyde Bishop, the day-shift foreman, passed through the gate as if I weren't standing there and climbed the steps to the back porch. Mom stopped him at the door. "I need to talk to Homer," he said grumpily. "Something's wrong with his phone."

"He isn't here," she snapped.

"Now, Elsie—"

"He isn't here for you, Clyde, and he's not going to be here. Don't try to call him either. I threw the black phones out in the yard, and that's where they're going to stay."

Mom drove Dad to the hospital the next day. I sat on the back steps to wait for their return. Dandy sought me out and, perhaps sensing I was sad about something, put his head on my knee. I stroked the soft blond fur on his head. Occasionally, he'd sigh as if he were deep in thought. After a while, Poteet joined us, sitting at my feet like she was guarding me. We were still there when my parents got back. Dad's head was bandaged and there was a thick patch over his eye. When he got out of the Buick, he had to lean against it to get his balance. Mom put his arm over her shoulder and took his weight. I rose to help, unlatching the gate for them. Even though she was struggling, Mom kept me away with a glare.

I watched my parents go into the house. I flinched when the screen door banged shut behind them, loud as a rifle shot. I wanted to follow but I couldn't. My feet seemed rooted to the yard. The sound of the screen door hung in the air, as if every door in Coalwood were slamming shut in my face, one after the other. My whole life, I had always been busy with some scheme to make things go my way. Now I knew there was nothing I could do to make things right, not now, not ever. At that realization, every ounce of energy in me just seemed to drain away. My arms hung limply at my side and I lowered my head in hopeless shame. Despite Doc's caution, I was about to feel sorrier for myself than even I could imagine. Then, as if some thief had sneaked up behind me and robbed me of everything I had always believed to be right and holy, I felt a terrible thing. It took me by surprise and I knew instantly it was wrong but I couldn't do anything about it. The boy who raised his head and looked around at the ugly old mountains surrounding him seemed far different than the boy I'd been merely moments before. Perhaps my lips didn't curl into a sneer but they could have. The worst thing I had ever felt in my life had taken control of me. I felt: *nothing*.

19

PICKING UP AND GOING ON

Auk XXI

IT WAS AS if somebody had reached up inside me and turned off a switch. I felt dull and thick. In the following days, I quit building rockets, quit studying my book, and quit going to the machine shop. I avoided all contact with my parents, never approaching Dad's bedroom, getting up early and standing in the dark an hour before the school bus came.

I was scared, but I wouldn't admit it, even to myself. Had I turned into one of them at last? Had Mr. Bykovski's death—my fault, no question—and Dad's accident finally got me where I was always supposed to be? Was I finally a good West Virginian, all stoic and stolid, filled everlastingly to my chin with guilt but not capable of showing it? I considered going to church, just walking inside and dropping to my knees in front of the cross, and begging for pain. Christ had felt pain—it was His gift to us, really. Weren't we, His people, supposed to be like Him? I detested myself for what I considered the abomination of feeling nothing.

Mr. Ferro found me walking past the church and made me stop and talk to him. "The boys got you another rocket all ready to go. How about you loading it and let's fire it off this weekend?"

"Tell them thanks, but I'm not in the rocket business anymore." I started up the steps of the church and then turned back. It was just another company building. I would have gone to see Little Richard, but it was too far and what was the use? He'd quote scriptures to me I already knew.

Jake! My heart leapt with the thought of him. Was there ever anybody who could feel more love of life than Jake Mosby? I ran to the Club House, but Mrs. Davenport said he had left for Ohio that morning.

I got on my bike and started to ride home, but Mr. Van Dyke hailed me from the steps of his office. I stopped and, my eyes downcast, listened while he praised Dad. "His courage should be an inspiration to us all," Mr. Van Dyke concluded.

"Yessir," I replied, the good Coalwood boy speaking by rote. "I'm sure you're right."

After the general superintendent was through with me, I tried to bike out of town central, but Mr. Dubonnet caught me outside the union hall. "Sonny, wait up," he called and then trotted up beside me when I finally stopped. "How're your parents doing?"

"Mom's fine," I said politely.

"And you?"

"I'm very well, sir." More of my learned rote. "I'm in a bit of a hurry."

Mr. Dubonnet held on to the handlebars. "I know what Ike meant to you, but you need to give yourself a break."

"I'm still in that hurry, sir."

He let go. "Then I guess you'd best be on your way, hadn't you?"

I lay in my bed at night and stared up into the darkness. When Daisy Mae purred, I petted her head, but didn't talk to her. I didn't have anything to say. I didn't pray either. I just waited for the night to end.

In the mornings, I looked at Coalwood and it looked nasty. The company was pulling up the train track, and the work crews were out, teams pulling the spikes out of the ties and hauling off the rails. The track bed left behind looked like an ugly black scar slashing through town. Without the coal cars, Coalwood was no longer covered by a daily pall of dust, but I still saw a gray, ugly crust on everything and everybody that would never come off.

I went to school now, unafraid of bad grades. I wasn't afraid of anything. Roy Lee made other kids move on the bus so he could sit behind me. "How's it going, sport?" he asked over my shoulder.

"Going," I said, and then closed my eyes and pretended sleep.

I sat alone in the auditorium before classes and at lunch and snarled at Quentin when he tried to sit beside me. "Stay away from me," I told him. He leapt up as if I'd kicked him.

I saw Valentine and Buck sitting together as if they had everything in the world to talk about. I didn't intrude. She had accomplished her act of compassion. She caught me looking and cocked her head, smiling tenderly. I looked away.

Dorothy and Jim walked down the hall toward me, hand in hand. Dorothy tried to talk to me. "Sonny—" She had to move or I would have walked right over her.

"What a dope," I heard Jim tell her.

Miss Riley made me stay after class. "Sonny, I'm sorry about your dad. How is he?"

"Fine." I waited for whatever she had to say to be said so I could leave.

She worried over me for a little while. "What's this I hear about you not working on your rockets anymore?"

To my surprise, I felt something. "That's right," I told her. My heart sort of hurt. I held my breath as if I were standing on ice that was cracking beneath me.

"Why?"

"Why not? Who cares?"

"I do. Quentin does. All the boys do."

"Then they can build their own rockets," I said with an arrogant snarl. "As if they could without me."

"You're feeling sorry for yourself," she said quietly. "And not a little bit proud. A poor combination."

My temper flared like a sweet electrical current coursing through my body. "What do you know about how I feel?" I indulged in my self-pity, the nastiness inside me curdling like putrid milk.

Miss Riley didn't even blink during my little tirade. "Give me your hand," she said.

"What?"

She reached out and took one of my hands, which I found I had balled into a fist. She unwrapped it. Her hand was soft and warm. I knew mine was cold. I had been freezing since the accident. No matter how I piled on the blankets at night, I was still cold. "Sonny," she said, "a lot has happened to you, probably more than I know. But I'm telling you, if you stop working on your rockets now, you'll regret it maybe for the rest of your life."

I pulled my hand away from her. I couldn't let her confuse me. I needed to stick to my new course. It was the only way I was going to get through all the mess I'd caused, make it right somehow.

"You've got to put all your hurt and anger aside so you can do your job," Miss Riley said.

There it was, the West Virginia thing—*the almighty job*. I should have known *that* was coming. Oh, yes, we all had our *job* to do in this state, breaking our backs to ship our wealth out to the world so we could turn around the next day and do it all over again for next to nothing. "What's my job?" I demanded harshly.

She ignored my tone. "Your job, Sonny, is to build your rockets."

"Why?"

"If for no other reason, because it honors you and this school."

I wanted to run from her, tear out of her classroom and keep going, and never look back. "What if I don't like doing my job?" I argued weakly.

She gave me a look that went down to my marrow. "Then, and especially then," she said, "you give it everything you've got."

SHERMAN called me. "Sonny, I think you'd better go up to the Little Store bus stop."

"How come?"

He told me. Then he told me what I should do. If Sherman said it, brave and good Sherman, I knew it was so. I burst for the door.

Mrs. Bykovski stood alone across the road from the Little Store with two cheap suitcases by her side. Mr. Van Dyke had given her a whole month in her house, but she had decided to leave after the standard two weeks. Sherman said she was going upstate to live with relatives near the hospital that housed her daughter. "I came to say how sorry I am," I told her. When she just looked at me and didn't say anything, I stood as straight and tall as I could and said, "It was my fault Mr. Bykovski was at the face."

I was startled when she smiled. "Ike could've gone back to the machine shop anytime he wanted. He said your dad got mad sometimes, but in the end he was always fair. The thing was, he didn't want to—I didn't want him to either. We got used to the money."

"But it was me—"

"Ike loved you, Sonny," she said. "You know that?"

"Yes, ma'am, but—"

"Shut up," she said evenly. "Just shut up." She sighed and looked up and down the street. "This was a nice place. Clean and peaceful. Wish we could've stayed."

My heart felt like somebody had put it in a vise.

The bus was coming up the road. "But things happen," she said.

I picked up her suitcases, helped the driver put them aboard. "Just don't forget Ike," she said at the door.

"I won't," I swore, and stood away from the bus.

She went to a seat and slid open the window and smiled down at me. "There's one thing you can do," she said. "Something I know Ike'd really like."

"Ma'am?"

"Keep firing off those rockets!" She closed the window and gave me a sad little smile and the bus pulled out. I looked up the road until it disappeared around the curve at the base of the mountain.

There was a breeze coming down the hollow. The dogwoods low on the mountain waved as if asking me to look at their glory. They were like white bouquets God had stuck in the stands of ancient oaks and hickories, glistening green in their own new growth. I heard something and looked up and down the road for its source. It wasn't just a single sound. It was Coalwood moving, talking, humming its eternal symphony of life, work, duty, and job. I stood alone on the side of the road and listened to my town play its industrial song.

Auk XXI was fired three weeks after the accident, about the time my father, ignoring Doc's orders, got up and went back to work, his right eye blind and glassy, a bloody rent full of stitches still on his forehead. Mom watched him go and then went back to her table and sat in front of her beach picture and pretended not to notice when the company men came inside and hooked the black phones back in. There wasn't anything for me to say to either one of them. We all had our jobs to do.

Mom fixed supper every day, but left it on the stove, retreating to her bedroom. Jim and I spooned out what we wanted and we took it to our respective rooms. Dad rarely came home from the mine, and when he did, he ate the cold leavings. I fully expected the rest of my life in Coalwood to go on exactly like this. Jim had his football scholarship and was leaving in July. I had one more year and then—whatever it took, I would go too. I did not plan on taking a dime from my parents for college or anything else. Jake had always said the Army or Air Force would like to have me. Joining up seemed a good idea. There was always the GI Bill. I'd get my college and get on down to Cape Canaveral in my own good time.

The seamless steel tubing I told Mr. Caton to order was delivered. He said Dad signed the request without comment. Quentin and I still weren't quite ready to work the equations for a De Laval nozzle in my book, but I got the machinists working, instead, on a new nozzle with deeper countersink cuts, hoping we might acquire at least some of the attributes of the converging–diverging design.

"Watch this rocket fly," I predicted to the boys at a morning BCMA meeting in the auditorium. "This one's really going to go!" I had apologized to them all for shunning them after the accident, and they had all acted as if nothing had ever happened. That was the West Virginia way, and they were better at it than I was.

From the moment I pressed the firing button the following weekend, I knew it was our best rocket yet. A hot, conical flame spurted from *Auk XXI*'s base as it accelerated off the pad. It darted for the blue sky, a long white contrail flowing out behind it. A couple of other Big Creek boys, Dean Crabtree and Ronnie Sizemore, had joined us for the day to help out around the range. Mr. Dubonnet, our machinists, and thirty other Coalwoodians cheered. Basil danced around his Edsel. The auk hurtled down-range, exactly as planned. I savored the *thunk* it made when it hit the slack. It was perfect. "Four thousand, one hundred feet," Quentin reported from his downrange post.

"We picked up a thousand feet," Roy Lee said. "We'll be at a mile next time."

"We will," I said, "but this is the last time we fly with rocket candy." I had already talked over what I was about to say with Quentin. "Next time we fly will be with a new propellant: zinc dust and sulfur. We're going for maximum altitude."

Sherman frowned. "What do we know about zinc dust and sulfur?"

"Not much, but we'll learn."

"But rocket candy's doing great!" Billy protested.

"Yeah," O'Dell weighed in. "I don't think we ought to change."

"Zinc dust and sulfur," I told them. "That's next. If you don't like it, quit."

"Who elected *you* king?" Sherman demanded.

"I'm in charge," I replied like the tough man I was trying to be, "and that's the way it is."

Roy Lee hung back while the other boys stalked off. "For God's sake, Sonny, take it easy."

"Don't start with me, Roy Lee," I told him. "Quentin and I are going to design a sophisticated nozzle, and we need an advanced propellant to go along with it."

"Okay. Then why didn't you explain that to the others?"

"Because I don't have time to explain everything I do."

"What's your hurry?"

"I—*we* are going to win the county science fair next year. And to do that, we're going to have to be twice as good as any of the Welch High students. There's a lot to be done—and learned—before then."

"Why do you want to win the science fair?"

"Do I have to defend everything I do? Don't I already do everything in this club anyway?"

Roy Lee looked grim. "No, you don't. But even if you did, I think you shouldn't talk to the boys that way."

"I don't give a rat's ass what you think," I said, biting off each word.

Roy Lee unleashed a surprise punch to my chest. I went sprawling onto the hard slack. I rubbed my chest—it hurt—while he stood over me, his fists at the ready. "You moron," he hissed. "We've worked our tails off on *your* rockets. So you think you can just come down here and treat us like we're nothing? If that's what you think, come on, get up. I'll knock you down again!"

I sat up on the slack, still holding my chest. "I just want to use zinc dust and sulfur," I said shakily.

"Jesus, what an idiot you are," Roy Lee said, shaking his head. "Use whatever the hell you want." He reached out his hand to help me up. I took it and he drew me to my feet. "I'm sorry I hit you."

"I'm not," I said, and I wasn't.

20

O'DELL'S
TREASURE

My dad's eye had not healed properly. He had kept it, but it was unfocused and watery. The doctor in Welch said it would likely remain that way for the rest of his life. Dad held one hand over his bad eye to read his newspapers and magazines and to watch television. He and Mom had made a kind of peace as far as I could tell. They acted as if nothing had happened, although they rarely spoke. Dad and I had little to say to one another. Mom spoke to me kindly, asking how schoolwork was going, but nothing of consequence. Jim might as well have been a ghost in the house as far as I was concerned. It was rare when the family had supper together, and when we did, there was only the lonely clinking of our forks and knives on our plates. There was at least Daisy Mae, ever my gentle confidante.

Mom stayed in her room in the morning and let me get up on my own. Without her prodding, Jim had no trouble waking up in time to spend his hour in the bathroom, but I missed the bus a couple of times, had to hitch a ride over, was late, and got called into Mr. Turner's office. If it happened one more time, he promised, I was in more trouble than I could possibly imagine. It happened again in mid-May, despite the alarm clock I borrowed from O'Dell. I found myself standing across the filling station with my thumb out toward War. Jake pulled up in his Corvette. "Where to?" he asked, swinging open the door. It was good to see him. He'd been in Ohio since the accident. I didn't know why.

"Big Creek! I'm late!" I clambered inside.

A gleam came to Jake's eyes. "Well, all right!" he cried and slammed his foot down on the accelerator. We shot past the mine and up Coalwood Mountain. He was drinking and handed over the bottle. "You drink, don'tcha?"

"Not before school," I replied, a true statement as far as it went.

He put the bottle to his lips and made three successive curves at once without looking. I saw Geneva Eggers in front of her house, wearing canvas trousers and a plaid shirt and sitting on a fence post. I slid down in my seat when Jake slowed the Corvette to a crawl and rolled down his window. "Hidy, Miss Eggers," he said, tipping an imaginary hat and sliding into the southern West Virginia vernacular. "How you feelin' this fine mornin'?"

"Why, I'm feelin' jus' fine, Jake," she said. She peered inside the car. "Who you got there? Oh!" She smiled prettily when she recognized me. "How you feelin', Sonny? Did you miss the bus again?"

I slid ever deeper in the seat. "Yes'm," I mumbled.

"Well, y'all take it easy," she said as we idled on by.

"Oh, that's a fine woman!" he said over the shriek of tires. "She sure seems to like *you*, old son. How do you know her?"

I shrugged. "Once, when it was snowing. She let me warm up by her stove."

Jake laughed a great open-mouth laugh as only he could.

When we got to Big Creek, I thanked Jake and made a run for chemistry class. When I got there, I realized I had left my books in his car. Miss Riley was sitting at her desk, placidly noting our arrivals, when Jake strolled in with them. She glanced at him a second or two longer than I thought was necessary and then went back to her attendance roster. I waved and he brought my books over. "Who's your teacher?" was all he wanted to know.

I introduced them and they shook hands. "So I believe you are an engineer, Mr. Mosby?" Miss Riley asked in a dulcet voice I had never heard her use before.

"I have a degree, Freida, but some would question if I am an engineer," he replied, oh so smoothly. "The rocket boys tell me you're their favorite teacher, and I must say I'm impressed by their taste."

She blushed and looked down at her papers. "Well, do call again, Mr. Mosby."

"Jake," he answered, all but flipping his eyebrows up and down in anticipation. "You can count on that, Freida."

He bent over me on the way out. "Forget the school bus, kid. You've got a chauffeur anytime you want it!"

I'm not certain when it dawned on me that Jake and Miss Riley were going out, but one day she called me to her desk after class and wanted to know more information about "Mr. Mosby," who his friends were, what people thought of him, and so forth. I lied, of course, and told her how respected and beloved he was in Coalwood. I figured I owed Jake at least a good recommendation. It was a character-building exercise, since, to my surprise, I was jealous.

As the school year wound down, the junior class and I walked the halls with a sense of impending ownership. We suspected that the good times were finally about to roll. The BCMA met in my room just before school let out. We had several issues to discuss. We needed money, I said, to buy zinc dust. We also still had a debt to Mr. Van Dyke for the telephone equipment.

O'Dell looked around to make certain no one was listening. Other than us boys, only Chipper and Daisy Mae were in the room, and both of them were asleep. Satisfied, he bade us to lean in closer. "There's scrap iron out there, boys," he whispered. "It's like gold. All we have to do is dig it up."

He explained. With the track pulled out of Coalwood, the N&W Railroad Company had abandoned the spur line in the wilderness of Big Branch, about five miles west of Cape Coalwood. Underneath the track were cast-iron drainage pipes. "Here's what we do," O'Dell said. "We dig up those pipes, bust 'em into pieces, and sell the lot to a scrap yard. We'll make a ton of money and it's all legal!"

"Why don't we just take the rails?" Roy Lee wanted to know in a burst of logic and suspicion. "They're easier to get at."

"If we took the rails, the scrap yard might be suspicious and tell the railroad company," O'Dell answered.

"What difference would that make?" Sherman wanted to know. "If it's legal to get the pipes, isn't it legal to get the rails?"

Sherman and Roy Lee's doubts did not fit O'Dell's universe. "No," he said, as if that explained everything.

It took a month to get our expedition prepared, so we were approaching the end of June by the time Red drove me, Sherman, O'Dell, and Roy Lee in the back of the garbage truck to the abandoned track. Quentin and Billy were visiting out-of-state relatives for the summer. O'Dell's father left us with a pile of assorted supplies, including a canvas tent, sleeping bags, a camp stove, four grocery bags filled with canned food (mostly beef stew), a couple crates of bottled sodas, some loaves of white bread, a couple of big cartons of Moon Pies, several boxes of matches, a wheelbarrow, two shovels, two sledgehammers, and a pick. The tools and the wheelbarrow were borrowed from several different families. We'd pooled our meager resources to buy the food. We set up camp in a small clearing and then went looking for our first pipe. We found it a hundred yards past a timber trestle bridge. We looked over the bank to where the pipe protruded. "Good God," Roy Lee moaned, "It's under ten feet of dirt!"

"What did you expect?" O'Dell demanded. "You think the railroad company was going to lay drainage pipes on top the ground?" He took a shovel and pushed it into the track bed. It penetrated barely a half inch. "It *is* kind of hard," he admitted.

I picked up the other shovel, and a copperhead snake crawled out from underneath it. I jumped about a foot straight up in the air and came down running. "Stand back," O'Dell said. "I'll take care of this."

He took a big swing at the snake with his shovel, missed the terrified reptile by at least six inches, lost his footing, and fell over the embankment. I ran to the edge and looked for him, but all I saw was a muddy stain in the river far below. I called out to him, but there was no response. Both Roy Lee and Sherman sat down on the track, holding their stomachs from laughing. "What if he broke his neck?" I worried.

"Take more than a hundred-foot fall to hurt that boy," Roy Lee said in all seriousness.

When O'Dell finally climbed back to the track bed, wet but unhurt, we began to dig. We dug all that day and the next and the next. It rained every day, and we found ourselves working up to our necks in sticky brown mud. At camp, everything became covered with mold. Our food started to rot.

I didn't care. After each day's labor, we'd wash in the river and sit around our camp fire and listen to the sounds of the dense forest—the wind rustling through the trees, the munching of deer feeding on crab apples, the crackle of raccoons in the underbrush, the owls hooting their mournful cry. After months of tension and worry, I felt grateful to be detached from Coalwood and all its problems. I hadn't realized how lonely and miserable I'd become until I went down into the wilds of Big Branch. There, with nobody around but Roy Lee, Sherman, and O'Dell, I could be just another boy again. I put Coalwood and even my parents out of my mind and took in all the sounds and sights and smells of God's nature everywhere about me. For the first time in months, I was genuinely happy.

When it got dark and a river of stars flowed overhead, we spread out our sleeping bags and laid down on them and talked. After we exhausted the topic of females, we considered the future. We agreed we were all going to make it into space. The United States would need men like us to be explorers and adventurers. It would be an adventure like the one we were having at Big Branch. If we lay there long enough, we were blessed with a satellite—American or Russian, we were never sure—streaking across the sky. It still made my heart beat a little faster to see one.

After five days of digging, we were finally rewarded with the sound of a pick hitting cast iron. When we finished uncovering the pipe, we looked like the mud men of Borneo worshiping a fallen idol. Sherman picked up a sledgehammer and fell to smacking the damn thing. I took the next turn at the hammer, smacking the pipe over and over until a crack appeared. O'Dell finally succeeded in knocking a chunk loose. We all jumped into the hole and admired the triangular iron fragment. It had taken nearly a week of hard labor to extract that tiny little piece. Undaunted, we got stronger every day and refined our technique. Red ferried in more food, and two weeks later, we had piled up a treasure trove of cast iron almost as high as our tent.

We were on our tenth pipe when I clambered into its deep hole to take my turn at breaking it up. After hammering out a pile of jagged iron fragments, I slipped while crawling out and threw back my left hand to catch myself. My hand plunged into the pile of iron, and I could feel a sharp edge slice into my wrist.

It didn't hurt that bad. Roy Lee started laughing, not because my fall was particularly funny, but because he was so exhausted he had turned giddy. I pulled my hand out of the cast-iron fragments and looked with wonder at my wrist, painted a bright red, and then I saw a spurt of blood. I laughed as another geyser of blood went flying. O'Dell saw it and laughed too, and so did Sherman. "Look," I giggled, crawling out of the hole, "I'm bleeding to death."

Roy Lee sat down, his face flushed from laughing. "You are," he chortled. "You really are!"

"Let me see that!" O'Dell said, sobering up. He held up my wrist. There was an inch-wide slit across it. I could see the *O* of the severed artery. Suddenly, I felt dizzy and sat down and stared dumbly at the scarlet geyser until I started to giggle again, and then we all giggled inanely at each other. "We got to stop the bleeding," O'Dell finally managed. He took off his T-shirt and pressed it against the cut, then tore a strip from it and made a tourniquet. He wrapped it six inches above the cut, using a stick to wrench it tight. "We got to get Doc," he said.

I was no longer laughing. As hot as that summer day was, I felt cold. I sank down on a rail and hung my head, my brain spinning. "I'll wait here while you get him," I said, my eyes suddenly heavy. "Maybe I'll just take a little nap."

"If we walk out, it'll take us most of the day," O'Dell said, checking the height of the sun, "and it'll be dark. By the time we make it back . . ." He looked at me. "Sonny, wake up! You've got to walk out with us."

I slid off the track and wallowed in the dirt, feeling the sun on my face. "Oh, I don't think so. . . ."

Sherman and Roy Lee started to laugh again, but O'Dell stopped them. "We got to get him to Doc, you guys. This is serious!" He cranked the tourniquet tighter. "We got to or he's going to die."

"Die?" I perked up. "Who's gonna die?"

"You are, you moron!" O'Dell said, and grabbed me under my armpits and tried to pull me to my feet. The other boys, sobered now, helped me up. I leaned on Sherman and we started to walk up the track.

It took us six hours to walk out. When we got to Frog Level, it was nearly dark. I lay down in the middle of the road while O'Dell went after his father. I saw a satellite and then another, streams of them dashing across the heavens, red and pink and white and blue and green, and then the bowl of sky began to turn, slowly at first, and then faster and faster. Sherman and Roy Lee took turns waking me up, but by the time O'Dell came back, I had passed out. Red picked me up and loaded me into the back of the garbage truck. At Doc's house, his wife appraised me at the front door, holding her nose at the smell, and announced that Doc wasn't there but was probably at his office. That's where we found him. He also held his nose (the back of the garbage truck had added a final eau de pig slop to the mildew and slime from our camp) and led me back to his examining room. He sat me on the table, unwrapped the tourniquet and the soggy dressing, observed with interest the waning spurt of the little blood left inside me, and hauled out his suture kit. "You want me to deaden your wrist?" he asked.

"Oh, yes, sir," I said groggily.

He shrugged. "Your dad got all his stitches in his forehead without any painkiller whatsoever."

I rose to his challenge. "Then I don't need it either."

Doc plunged the needle in. It hurt and I howled. "Deaden it. Deaden it!"

"Naw, too late now," Doc said. He happily sewed while sweat popped out in huge drops on my forehead. I swayed, nearly passing out every time he plunged the needle in. "Very good, Sonny," he said after a lifetime. "All finished. Hop up."

I hopped up and fainted. I woke on the cot, with Mom looking down at me. She had her nose covered with a handkerchief. "Sonny, dear God but you had me worried."

"Hi, Mom," I smiled weakly.

Doc's face appeared. "The other boys walked him out. Their tourniquet pretty well saved him." He lifted my arm, checked the bandage around my wrist. "You're going to have a nice scar, young man, to remind you of this particular adventure."

"Can he go home, Doc?" Mom asked.

"I hope so," he said. "The fumigators are due any minute."

Mom walked me out, the other boys jumping to their feet in the waiting room. They clustered around the car as I sagged onto the bench seat. "Go get the scrap iron," I gasped at them.

Mom made me take a shower in the basement before letting me upstairs. I lay down on my bed and heard Dad come in from the mine. Then my door swung open. He and Mom entered. "You okay, little man?"

It was wonderful to hear Dad's voice. "I'm fine," I said. I looked up at my parents. It made me so happy to see them together, it took everything I could muster not to cry. "I'm sorry. Like always."

Dad said, "You have nothing to be—" but I fell into a deep sleep before he finished. It was a sleep packed with dreams of swirling colors, as if I were in the midst of a gigantic kaleidoscopic whirlpool. Once I came to and was surprised to see Jim in the bedroom with me, sitting at my desk watching me with something like concern on his face. Was I dreaming?

The next day, while I still slept, the other boys went back to Big Branch and loaded the pile of scrap iron on the garbage truck, forgetting the tools and the wheelbarrow, and carried it to Welch and the Chester Matney Scrap Yard. Mr. Matney weighed it carefully—more than four hundred pounds!—and then counted out the twenty-two dollars and fifty cents he figured he owed us. We had expected at least a dollar a pound. No, Mr. Matney, said, prices were down. After subtracting out the cost of the food and ignoring the cost of the ruined sleeping bags and the lost tools, we made a grand total of four dollars. That was before Doc sent me a personal bill for five dollars for "sutures and labor."

Jake came to our rescue. If we promised to wash and wax his Corvette for approximately the rest of time, he'd cover our debts. We took him up on it, paid off Mr. Van Dyke and Doc, plus the people whose tools we'd lost.

We also bought ten pounds of zinc dust.

21

ZINCOSHINE

Auks XXII, A, B, C, and D

NOW WAS THE time for greater strides. With Quentin and Billy back in the fall, we gathered at Cape Coalwood to test the first of our zinc–sulfur rockets. Our machinists were there, along with Reverend Richard and at least a hundred other Coalwood citizens. The Reverend beckoned me over and took off his hat. "Been prayin' for you, boy," he said. "Knew you needed it."

He told me then that he had had a dream. He had seen men on the moon, and I was one of them. When he woke, he had opened his Bible and his eyes had fallen on the testament of Peter. *"Nevertheless we, according to His promise, look for new heavens and a new earth, wherein dwelleth righteousness,"* he quoted.

"I hope that's so, sir," I told him, and he looked pleased.

We loaded the rocket on the pad by holding it upside down, tapping the casement with a hammer to get the gray powder to settle as we poured it in. When I pushed the ignition button, *Auk XXII* detonated on the launchpad, sending shards of steel deep into the slack and a boil of greenish-white smoke into the air. Had we already passed the critical dimension for our new propellant? I couldn't believe it. There was something else wrong.

The crowd wandered over to help us study the fragments like tea leaves. "Guess the Lord didn't want this one to fly," Reverend Richard said while I turned a piece of jagged steel over and over in my hands, looking for some clue as to what had happened.

When our audience drove away, we had a BCMA meeting. Our

conclusion was we didn't know what we were doing with the new propellant. No fingers were pointed. It was just the way it was.

I went home that night, unhappy and confused. Mom gave me a look as I crossed the kitchen. I knew she had something to say. "What?" I asked her.

"The company's selling the houses," she said.

IN the fall of 1959, the television and newspapers were filled with stories about American rockets roaring successfully into the sky. *A-OK! We have lift-off! All systems go!* The language of our rocket engineers became part of our everyday speech.

On September 9, 1959, I read that NASA had launched a mock-up of a *Mercury* manned capsule aboard a *Big Joe* rocket. It was a suborbital lob shot, and though the spacecraft was empty, the newspapers said it was the opening round of America's program to put men into space. I was thrilled. I started to think about space-craft big enough for whole families to be launched—to the moon, Mars, perhaps the stars themselves. As Reverend Richard had said, to look for new heavens and a new earth. I often felt a new earth for me would be an excellent plan.

Most people in Coalwood were preoccupied with other matters more down to this earth. Not only had the steel company decided to sell the houses, the sewage and water systems and the churches were on the block as well. Would the mine itself be next? Mr. Van Dyke and Dad went to the union hall and sat around the table to inform the union leaders about the sell-off. I listened from the dining room while Dad in the kitchen gave Mom the blow by blow of that meeting. "Dubonnet was all over me, asking how the men were going to pay for their houses and everything else," Dad said.

"Well, how are they?" Mom wondered.

"The company's going to loan them the money, almost no inter-est rate, twenty years to pay."

"That's still going to be money out of their pocket," Mom pointed out, "and then when they've bought the house in twenty years, who's going to buy it from them when they want to retire and move? Did you ever stop to think the reason the company wants to sell everything is because they don't think the mine has a future either? That's what the people are saying."

He used his good eye to look at her suspiciously. "Who've you been talking to?"

"It's common knowledge."

"Common knowledge is wrong," he growled. "The company's selling the houses because they hired some fool efficiency expert who told them it's cheaper to sell them than keep them up. Captain Laird knew that thirty years ago. It wasn't efficiency he was after. He said if the miners lived in company houses, they'd feel like they were part of the company, be more loyal to it. Anyway, Van Dyke's going to drive up to Ohio and tell them this isn't such a good idea. We'd like to keep things in Coalwood just as they are. I think he'll be back with good news."

"Oh, Homer," Mom said in despair.

A week later, Mom told me that Mr. Van Dyke had been fired. Mom guessed it was because he had lodged his complaint about the property sale but she wasn't certain. In any case, a new general superintendent was being sent to Coalwood to make sure the sale of the houses and the churches and the utilities stayed on track. Dad retreated again to the mine. The union hall rattled with outrage. I rode by on my bike going to the machine shop and heard the chant: *"Strike, strike, strike!"*

THE zinc–sulfur powder mix was too loose, I decided, probably because it had air pockets in it. That's why our last rocket had exploded. It was the same experience we'd had with black powder and then again with potassium nitrate and sugar. Before, a binding agent had solved the problem with black powder, and melting had produced our reliable rocket candy. Melting zinc and sulfur was not a good plan, I didn't think—it would surely detonate before the melting began—so I used dextrose and water to see if that would work as a binder. The resulting mix just spewed feebly in the hot-water heater. I wasn't sure why. "The water probably caused the zinc to oxidize," Quentin said.

"How about we mix in gasoline?" O'Dell proposed.

"Too dangerous," Quentin said. "And I'm not so certain gasoline wouldn't react with the zinc in any case."

We argued various liquids. Naphalene? The mine had a lot of the solvent derived from coal tar, but it was too volatile. Diesel? Not volatile enough. Semisolids such as paraffin? Too messy. Billy proposed alcohol, and Quentin's eyes lit up. "Yes! Alcohol is stable and it'll evaporate quickly. Perfect!"

We looked at scrounge-master O'Dell, and he grinned. There was only one place to go in Coalwood when you wanted to buy one hundred percent pure two-hundred-proof alcohol.

''ARE you sure Tag's not going to catch us?" I nervously asked again from the backseat of Roy Lee's car as we bounced up the rutted dirt road to Snakeroot. Roy Lee, O'Dell, and I were on a quest that had seemed like a good idea when I'd first heard it. Now I was having some doubts.

"Tag couldn't catch a cold," Roy Lee said, gripping the steering wheel and turning it back and forth to dodge the pot holes.

"He caught you and O'Dell in the mule barn."

Roy Lee shrugged. "That was different. This is a time-honored tradition. Every boy in town goes to John Eye's, sooner or later."

Because it was a Friday night, Roy Lee had to search for a parking place. A line of cars had already collected up and down the ditch line in front of John Eye Blevins's moonshine palace. I nervously handled the four dollars we had scraped together. Four dollars would buy us a gallon of John Eye's finest. We waited covertly in the shadows until there was a gap in the traffic and then clumped up the old wooden steps, worn smooth by years of traffic. A little girl with her hair tied up in braids swung on a porch swing and watched us with wide round eyes. "Y'all be too young to buy 'shine," she pronounced.

"Who are you, the moonshine police?" Roy Lee asked.

"Nuh uh." She shook her head. "I'm just smarter'n you."

A huge bulk suddenly filled the door. "Whatchall boys want?" The voice seemed to rise from a deep well. I held up the money. "Get on in here!"

The little living room had a broken-down couch and some easy chairs with the stuffing hanging out. A big old radio filled a dark corner, and on top of it was a cheap suitcase phonograph. Music I'd never heard before was playing, some sort of jazz. Hanging beads marked the door to the kitchen beyond, where three Negro men sat around the kitchen table. They were playing cards and ignoring the business in the living room. John Eye looked at us with his brow furrowed, as if he were trying to make up his mind about something, and then he put out his huge paw, palm up. I counted out our dollars and he nodded and limped through the beads. I knew John Eye's legend—how the timber had given way in his section and how he held up the roof with his broad back until the other miners got out. When the roof finally caved, a piece of slate had cut off his foot at the ankle. That was the reason the company—meaning my dad—allowed him to make a living from his still on the ridge behind his house.

John Eye came out with four fruit jars of clear liquid. Roy Lee expertly held one of them up to the light. "There's no water in here, is there?"

"I don't cut my whiskey!" John Eye rumbled. "It's pure 'n bony-fidy. Wanta sip?"

Roy Lee brightened. "Yeah!"

"I don't think we should," I said quickly. "It isn't for drinking anyway. It's for scientific purposes." I flinched. I hadn't meant to confess that.

John Eye held on to the other quart jars. "Whatchall mean? Y'all ain't gonna drink my stuff? This is the best corn likker in the county! It'd be sacker-lijus not to drink it!"

"Aw, he's just kidding, John Eye," Roy Lee said, and pulled me off to a corner. "All these guys got razors on 'em," he whispered. "We got to be nice to them or they'll cut our throats. And we got to make sure this stuff is pure, don't we?"

"Well . . ."

"You've never had a drink, have you?"

"Not exactly."

Roy Lee cocked an eyebrow at me. "Man, you think Wernher von Braun don't drink down at Cape Canaveral? I bet that's about all those rocket men do in between shooting off missiles. That and chasing women."

I couldn't resist his logic. I nodded agreement. "A-OK, John Eye!" Roy Lee told him.

Our host beamed and disappeared back into the kitchen, and I heard a cabinet squeak open and the tinkling sound of glasses being set up. The men at the table looked up from their cards and laughed. John Eye brought in a tray with three filled shot glasses on it. He held it out to us and we each daintily took the 'shine. Roy Lee held his glass up for a toast. "To Wernher von Braun!" He downed the drink, smacked his lips, rolled his eyes, and croaked, "Damn, that's good!"

O'Dell followed suit and wiped his mouth, tears streaming down his cheeks. "Good!" he shouted, but it came out a strained whisper.

Everyone looked at me. It *was* for Wernher von Braun, after all. I tossed back the drink, not even letting the liquid roll across my tongue. It went straight back to my gullet and caught on fire. I nearly doubled over as I felt it burn all the way to my stomach. I tried to breathe, but nothing worked. Roy Lee pounded on my back. "How about that, old son? Is that rocket fuel or what?"

"All . . . systems . . . *go!*" I finally wheezed.

"Wanta 'nother?" John Eye grinned, gold teeth flashing.

We boys looked at each other and then held out our glasses. "To Werrer va Brah!" we bellowed, while O'Dell and I sat down hard. Now I knew why all the chairs were busted in John Eye's living room.

Sometime later, we were singing "Blueberry Hill" as Roy Lee guided us uncertainly down the dirt road from Snakeroot. Tag pulled us over the moment we bumped up on the asphalt of the road that led down past the church. He fanned the sweet 'shine vapors away when Roy Lee rolled down the window. "Why, hidy, Roy Lee. O'Dell. Sonny. Whatchall boys doin'?"

An hour and a half later, I stood in front of my mother in the kitchen. I was wobbly on my feet and had a sickly grin plastered on my face. "You're drunk?" she asked in disbelief.

I didn't know if I was drunk or not but I was definitely sick. I had just spent an hour bowing before a ditch with O'Dell. Tag drove each of us home in turn. Punishment, swift and sure, was left to the family. Roy Lee's mom reached up and grabbed Roy Lee by the ear and dragged him inside the house. Red came out onto the porch, heard the story, and pointed at the shed in the backyard. The last I saw of O'Dell was him going down the path, head down, his dad close behind.

While Mom contemplated me, I held with crossed arms to my chest the sack with our four quarts of precious 'shine. Tag had let us keep it when we had explained its real purpose. Amazingly, he'd believed us. "Gawdalmighty, boys," he cried. "Whyn't you tell me? I'd of bought you some!"

I wasn't so drunk or sick that I had forgotten how to manipulate Mom. "Mom, I'm really, really, *really* sorry!"

She laughed. "Not this time, buster. There's no way you're going to wash the dishes or clean up your room or cook kidneys or anything else to get out of it. You're going to take John Eye's firewater down to the basement—oh, I know it's for your rockets—and store it with all the other crazy stuff you've got down there that could probably blow this house to kingdom come. Then you're going to go upstairs and take a shower and brush your teeth and get the smell of liquor off you and then you're going to bed."

"That's it?"

"That's it until I figure how I can do some real damage to you," she said with relish.

"But I've got to get this over with!" I cried.

"Well, that's just tough, isn't it? Now get out of my sight. I never could stand a drunk."

"Mom . . ." I was whining. "Hit me or think up something I got to do!"

She shook her head, smiling. "Nope."

I sunk to my knees and put my head in her lap. "I'm sorry," I said into the folds of her dress. "I'm sorry. I'm sorry. I'm *sorry*."

Mom touched my hair and then chuckled. "Sonny, I'm not going to punish you at all," she said. I looked up, stifling a smile. She saw the look on my face. "You little brat. I ought to hit you with a two-by-four. But promise me you'll never drink that stuff again and we'll call it square. Now, take it into the basement and then do what I told you."

I stood up stiffly, nodded, and went down into the basement with my white lightning. When Dad got home, he saw the jars and recognized them for what they were. He was in the kitchen with Mom when I came out into the hall after worshiping the toilet bowl again. "Elsie, did you know—"

"Yes, Homer."

"You plan on providing him an olive for his martini next?"

"We'll see."

God bless my mother, I thought, and went right back inside the bathroom. Even with my head inside porcelain, I heard her and Dad laughing. It was the first time I'd heard either one of them laugh for a very long time. I wished I could have joined them.

THE temporary general superintendent sent down by the steel mill in Youngstown was a man named Fuller. He talked like a machine gun and had about as much charm. He didn't move into the turreted house on the hill, but took up lodging in the Club House, reinforcing his status as a temporary fixture. He called in Mr. Dubonnet and the other union leaders and told them that fair terms were to be given, but the houses were going up for sale immediately and if anyone didn't like it, he'd have to get out. Fuller dared the UMWA to go on strike, waving the last union agreement. "There's nothing in here that says we have to provide you a house, water, electricity, or anything else. Any man says different is a damn fool idiot."

Mr. Dubonnet had to back down, and the sale proceeded. A utility out of Bluefield snapped up the sewer and the water lines, and within a month, the people in Coalwood started receiving bills for what they had always thought was naturally free. FOR SALE signs also went up in front of the churches. It felt, I heard people say at the Big Store, as if the company had chased even God out of town.

Reverend Richard and his congregation scraped up enough

money to buy his church and save it. Reverend Lanier, however, lost his job when the Methodists took up the mortgage on his church. Although the Reverend was a Methodist too, he had been a salary man for the company, and the denomination considered him tainted and wanted no part of him. The poor man had to pack up and leave. Mom said he went to California. I later heard he was on the radio out there. While the Methodists cast around for someone willing to come and serve in the wilds of West Virginia, the Coalwood Community Church, for the first time anyone could remember, was padlocked.

Then things really got bad. Mr. Fuller ordered a big cutoff and many men were given their pink slips. Mr. Dubonnet called an emergency meeting at the union hall and invited the company to explain. Mr. Fuller didn't show. I heard Dad tell Mom he had been ordered not to go either. Mr. Dubonnet came once more to our house. Dad let him inside, and their fight was loud and long. I listened from my basement lab as they shouted at each other for nearly an hour, neither man giving an inch. I heard Mom finally go into the living room and order them both out. "I just won't abide this yelling in my house, Homer, John," she said. They were still arguing as they went out the door, bound for who-knew-where.

I just kept working on my rockets. I mixed the moonshine into a zinc-and-sulfur compound and was rewarded with a thick gray composite that could be shaped like modeling clay. I squeezed it inside a toilet-paper roll and let it dry under the hot-water heater until the next weekend. With the other boys in attendance, I tossed the tube in the fire and had the door half closed when it went off with a huge *whoooosh!* The velocity of the burn was such that it flung the door open, blew the stovepipe loose, and sent white smoke gushing into the yard, along with us and the dogs. If that wasn't bad enough, when I went inside the kitchen, smoke had billowed through there too. I raced through the house, throwing open every window. The boys followed me inside and started flapping magazines and towels to clear it out.

Mom and Dad weren't home. Dad was at the mine. Mom was shopping in Welch. Men hanging around the filling station jumped the fence and came running, sure the house was on fire. Someone yelled that they had called the fire department in Welch. "Call back and stop them!" I pleaded. "There's no fire! It's okay!"

I heard stomping on the back porch, and Tag came inside followed by a short man shaped like a powder keg. I hadn't met him, but I knew who he was. "What the hell is going on here?" Mr. Fuller demanded, a stogie between his clenched teeth.

"Aw, it's just them rocket boys," Tag said, grinning. "You boys didn't set your white lightning on fire, didja?"

"White lightning?" Mr. Fuller shifted his cigar, giving me the evil eye.

With the other boys still fanning smoke out the windows, I quickly explained to the new general superintendent about our rocket-building and what the alcohol was for. He scowled. "Where do you fire these things?"

"Way out of town," I assured him. "A long way away."

"Still on company property?"

"Just barely," I said, wincing. His disapproval was made clear by the sour expression on his face.

Mr. Fuller turned on his heel and left, with the constable close behind. When Mom got home, she wrinkled her nose at the sulfurous odor, then went into the basement and looked at the ruined hot-water heater. Quentin and I came down and stood beside her, waiting for the wrath this time I knew I wouldn't escape. Her shoulders were shaking—I thought from crying—but then I saw she was actually laughing. She put her arm around my waist and then pulled in Quentin too. "You boys really are the light of my life," she said. "I've been wanting to get rid of this old coal-fired thing for years. I'm going to make Homer get me an electric one now. Why, I'll just open up the spigot anytime I like and there'll be all the hot water in the world. Just like the Rockefellers!"

Quentin stopped at the gate as he was leaving. "You have the greatest mom in the world," he said.

I looked over my shoulder. "She's got her ways," I said. I was just hoping she'd use them on Dad when he got home and saw what we had done.

Although I'm certain Mr. Fuller filled his ear about it, Dad made no comment on our abortive test. Junior from the Big Store arrived the next day with an electric hot-water heater. It was installed and humming happily as I loaded *Auk XXII-A* on top of the washing machine. I had decided to do the loading of our new propellant in sections, a few inches at a time. I pushed the zinc–sulfur–moonshine gumbo I had dubbed *zincoshine* into the casement with a broom handle. After each section, I placed the open end of the casement toward a fan to dry. After three hours of drying, I loaded another section. It was slow going, but after a week I had the rocket ready. We put up our posters and Basil wrote us up in his paper. Over two hundred people showed up the next weekend. The word was out that either we were going to have a great flight or we were going to blow the hell out of Cape Coalwood, just as we had Elsie's hot-water heater. Either way it was going to be a great show.

Auk XXII-A didn't disappoint. It leapt with savage energy from the pad, zinged up the guide pole, and split the valley with its thunder. The crowd backed away and *ahhed* as the rocket disappeared at the top of a tremendous column of white, boiling smoke. Quentin spilled out of the blockhouse and aimed his theodolite skyward. "Where is it?" he screamed. "I can't see it."

None of us could. It had flown out of sight, its huge smoke trail abruptly terminated. A little belatedly, I started to worry over where it might land. I looked at our audience. "Get in your cars!" I yelled at them, waving my hands. Some of them waved back.

"Time!" Quentin yelled out.

"I've got it!" Sherman replied, looking at his father's watch, which he had borrowed.

I kept searching the sky for any sign of the rocket, but I knew with my poor eyes I wasn't likely to be the one who saw it first. The crowd across from us was likewise engaged in searching the skies. I was starting to sweat. Where was it? Billy spotted it first. "There!" he yelled, pointing. I looked, but I still couldn't see it. Then I heard it. It was whistling as it came in. It sounded as if it were coming right on top of us. We ducked back into the blockhouse. A tree cracked behind us, and then there was the unmistakable sound of steel slamming into packed mountain earth. *Whump!*

"Thirty-eight seconds!" Sherman said.

Roy Lee looked at Sherman. "What good is knowing the time?"

Sherman explained our new calculations to Roy Lee. Quentin and Sherman and I had discussed it after Miss Riley had begun to teach us a little Newtonian physics. A falling body accelerated toward the earth at thirty-two feet per second. The equation for finding how far it fell was $S = \frac{1}{2}at^2$, or sixteen times the square of the time it took to fall. If one assumed a rocket took approximately the same time to reach altitude as it did to come back down—a good assumption, since zincoshine burned so fast our rockets were essentially free flyers the moment they left the pad—then dividing the total time of flight in half, squaring it, and multiplying the result by sixteen gave us a rough estimate of the altitude reached. Half of thirty-eight was nineteen, the square of nineteen was 361, and sixteen times that was . . . "Five thousand, seven hundred and seventy-six feet!" Sherman announced gleefully.

We had done it! We had broken the mile barrier! Billy had run after the rocket and came racing across the creek holding it high over his head. The crowd cheered from the road while we danced. "A mile! A mile! We flew a mile!"

"We're going into space," Quentin said after we stopped to take a breath. "We're really going."

"I've been meaning to talk to you all about that," I said. I gathered the boys in close. "I read where space begins at thirty miles. I think we can reach it."

At that moment, everybody was taken with the idea, even Roy Lee. "Let's do it!" he roared at the sky.

"Prodigious!" Quentin bellowed back. "We'll be on the cover of *Life* magazine for sure!"

After we'd calmed down a little, we cleaned up around the blockhouse and carried our rocket and our other gear to Roy Lee's car. I saw the last truck of the observers leaving. It was a company truck, and the man driving it was Mr. Fuller.

The silvery cylinder burst forth in a fiery column of smoke and flame, racing the very wind as it soared into the sky, a messenger of these boys of Big Creek, these boys who use their brains, not brawn, who play not football but with Apollo's fire. Oh, fleet rocket, your

thunder wanders down the valleys, startling deer and mountaineer
alike. How high? the crowd cries. How high will it fly? The boys
race from their bunker and go running down their slack-dump firing
range, the joy of youth and scientific interest playing across their
delighted faces. Oh, Rocket Boys, oh, Rocket Boys, how sweet thy
missile's delight against the pale blue sky. A mile, a mile, they cry.
We've flown a mile!

The McDowell County Banner, *October 1959*

To take our next steps, we needed to work the nozzle equations
in my book. But we were missing an important number—the specific impulse of zincoshine. Specific impulse was, my book said, the
thrust a rocket produced when it burned a pound of fuel in one
second. To make this calculation, we needed a way to hold a rocket
in place while we measured its thrust. O'Dell talked Mr. Fields, the
butcher at the Big Store, into loaning us a scale—the kind that
hung from a ceiling and was used to weigh sides of beef—with the
promise that it would be returned without so much as a scratch.

At Cape Coalwood, we clamped the scale to the underside of a
plank that sat on two sawhorses. Then we ran a wire from the
hook of the scale to the backside of one of our rockets, which lay
inside a slightly wider tube, which was in turn clamped to the plank
with a piece of fashioned strap steel. The idea, which was Quentin's, was that the rocket would move down the tube when ignited
and be tugged to a stop by a cable attached to a spring, which was
attached to a vertical stake in the ground. We'd watch the spring
with binoculars and see how much thrust was produced and then
divide the answer by the pounds of propellant burned and how
long it took. The result would be the specific-impulse number.

We had a good idea, but we got a bad result, a situation not
totally unknown to the BCMA. When we lit the rocket (we called it
Auk XXII-B), it threw out a plume of flame and smoke and zipped
down the tube, jerked the plank off the sawhorses, dived into the
slack, bounced once, and then, with unerring accuracy, turned in
our direction. It whizzed over the blockhouse, the scale on its stake
following on the cable, bounced off a rock in the creek, and then

flew up into the woods, where it was stopped by a hornet's nest. The hornets, not pleased, chased us up the mountain on the other side of the slack. We crouched there and watched as they marauded up and down the slack like a tornado. It was nearly dusk before they finally dissipated. We found the scale in pieces. The BCMA spent the next four Saturday afternoons cleaning up the butcher shop to work off our debt. Mr. Fuller came up from his office and watched us. He didn't say anything, just watched, and we knew he knew what we had done.

Our next attempt at measuring thrust was a variation on the theme, this time using Mom's bathroom scale. I was absolutely certain that I would return it unscathed, because this time we had a better test-fixture design (it happened to be mine). Using some channel iron found behind the machine shop, we built what looked like a miniature oil derrick and placed it over the scale as a brace for a tube that held the rocket. We positioned a mirror on the "derrick" so that we could read the scale with binoculars. We put the rocket, this one named *Auk XXII-C,* nose-down inside the tube and ignited it. The scale and the concrete pad was in its way, so all it could do was push. I knew it would work, and it did. In the first seconds of the firing, we were able to read the scale. Then, unfortunately, *Auk XXII-C* became a rocket jackhammer. It bounced up and down in its tube, attacking the scale. The scale survived the first few impacts, but then we heard a big *sproinggg* and it flew apart. When the propellant hissed down to completion, Mom's bathroom scale was scattered all over the pad.

I tried to put the scale back together as best I could, managing to at least fit all the parts back inside and pound the frame back into a rough semblance of its previous shape. I put it back where I found it, hoping Mom wouldn't notice. The toilet was still flushing when Mom banged open the door to my room. "I want a new scale," she announced. "And I want it within one day."

O'Dell came up with the new bathroom scale for Mom. I didn't know where he got it and I didn't ask him. I just placed it in the bathroom and backed out.

We'd gotten ourselves into trouble—no surprise by then—but

we also had a shiny new working specific-impulse number for zincoshine. Almost anything was possible now.

THE flight of *Auk XXII-D* was to be the final flight in the series that used countersunk nozzles. I had the machinists make just one minor adjustment with *Auk XXII-D,* to shrink the size of its fins. I had noticed our old rocket-candy rockets wobbling when they broke the plane of the mountains and got caught in the wind stream coming across the ridgeline. Smaller fins, I thought, would keep that from happening and make them fly truer. What I didn't know was that O'Dell and Sherman, who had taken on the job of setting our rockets up on the pad, had also noted the wind's effect on flight. They had been compensating for it by aiming our rockets at a small angle against whatever direction the wind was coming. It was, as it happened, a very windy day when *Auk XXII-D* made its flight. The wind was whistling across the slack dump. I studied the clouds. They were scudding along in the same direction, toward the west away from the center of Coalwood.

In an attempt to compensate, O'Dell and Sherman leaned the launch rod more than usual against the wind. I was too busy getting the blockhouse organized to notice what they were doing. I ran our ignition wire, made sure Quentin and Billy had set up their theodolites, checked with Roy Lee that our new fancy-looking firing console was in working order, and then called for the boys to go to their stations.

The word had spread after our last launch that our big new rockets were potentially dangerous to spectators. It didn't keep them from showing up, but I saw some of the miners had brought their helmets with them. Roy Lee ran our BCMA flag up the pole. That was the final signal. I looked at the flag with some concern. It was snapping in the wind—*pop, pop, pop.*

I put my worry aside. This was a big, heavy rocket. It should fly straight and true. I went inside the blockhouse and knelt behind the wooden console Sherman and O'Dell had built. They had even installed a switch salvaged off an old electric-train transformer. I toggled it, and the zincoshine-propelled rocket blasted off with a savage roar. It flew straight, slicing without flinching through the plane of the mountaintops. Sherman counted. "Ten, eleven, twelve . . ."

I watched as the smoke trailed off from the *Auk,* and then it disappeared—still on a heading for Coalwood. "No!" I yelled, horrified.

Sherman looked up from his watch. "Huh?"

Roy Lee saw what I was seeing. "Oh, *shit!*"

All of the other boys looked up the valley, and so did our audience. In unison, almost everybody repeated exactly what Roy Lee had just said. We ran for his car, the crowd scattering before us. We roared up the road, plastering our faces to the car windows, looking for any sign of our wayward *Auk.* "You know," Quentin said learnedly, "I suspect the velocity of a rocket also affects its stability at moments of maximum stress—"

"Shut up, Quentin."

"Perhaps there are ratios of wind pressure to velocity that can be mathematically calculated. Interesting! I think—"

"Shut up, Quentin!"

Frog Level was peaceful, so we kept going. Up ahead, at Middletown, we saw a crowd near the road. People were running to see whatever it was. Roy Lee moaned. "We've killed somebody!"

Auk XXII-D had landed in the field beside Little Richard's church, a patch of flat ground that was often used for pickup touch football and softball games. Totally panicked, certain we'd killed some unsuspecting playgrounder, we pushed through the crowd and found our rocket buried in the grass, with only its fins and nozzle showing. The odor of sulfur was strong. I was so relieved I started to laugh, and everybody in the crowd laughed with us. "Gawd, you boys will be flyin' all the way up to Washington Dee Cee you keep at it this way," Tom Tickle roared. Good old Tom was a miner who had always supported us.

Roy Lee got out a shovel from the trunk of his car and began to dig the rocket out, while people talked about where each of them was when it fell and what it sounded like and how it shook the ground. Mr. Fuller showed up in his truck. He took one look. "You boys are a damn menace!" he announced. "You've fired off your last rocket in this town."

Here we go again, I thought.

"Now, looky here," the Reverend Richard said to Mr. Fuller, "don't be talkin' 'bout stoppin' these boys. We're proud of 'em!"

"You can't shut our rocket boys down," Tom said. "Leave 'em alone. Weren't nobody hurt!"

A grumble of assent rose from the crowd. Mr. Fuller surveyed us. "They fly on company land, and from what I've observed, they do it with company property. I'm the company, and I'm telling you they won't fly anymore."

"Suh," Reverend Richard said, "you may be the comp'ny, but these men and these ladies around you, they is *the town.*"

"The boys keep flyin'," Tom said, stepping up to Mr. Fuller. "You can sell our houses, charge us for the air we breathe, I guess, but you ain't stoppin' these boys."

A woman I recognized as Reverend Richard's piano player jostled Mr. Fuller. "Our town was just fine before you came," she told him. "Now, don't you go pullin' those Yankee airs around here, mister, sayin' what you gonna do and what you ain't gonna do."

Mr. Fuller retreated to his truck. He stuck out his jaw in my direction. "I'll be talking to your dad!"

"Keep flyin', boys," Tom said to us. "Keep on puttin' those rockets in the sky."

"But it might be nice if you aimed a little more that way," somebody else said, pointing back to Cape Coalwood.

As soon as I got home, Dad called me up to the mine. I took a deep breath and then trudged up the path. The door to his office was ajar. He was pondering a mine diagram with one hand held over his bad eye. He turned at my knock, and when I saw him I was shocked at his appearance—not so much his watery ruined eye, but his gaunt face. I had seen little of him since the announcement that the houses were to be sold. He reached for his hat. "Let's go for a ride," he said, and walked me to his truck. I noticed he had a bit of a limp, although the accident had not, to my knowledge, hurt his legs. He seemed somehow smaller than I remembered him being.

I wanted to ask him where we were going and why, but I held back. He drove us slowly down the road, past our house and the Coalwood School and through the corridor of houses that led to the center of town. I was startled to see that one of the houses across the creek was in the process of being painted a bright yellow. In the line of houses still painted company white, it stood out, a bright, brave statement that things weren't the same anymore. It was a sold house, one of the first, and Dad looked at it and wiped his mouth as if he were throwing away the words he wanted to say.

We drove past the community church, the cross on top slightly askew. The padlock on its double doors was huge, more suitable for the gates of heaven or hell, I thought, than a little West Virginia church. We drove past Little Richard's church and then down through Frog Level. It was clear we were headed for Cape Coalwood. "I thought you should see this, Sonny," Dad said when I looked at him questioningly, "rather than just hear about it."

When Cape Coalwood came into view, I saw a bulldozer covering up our launchpad. Mr. Fuller was walking beside the bulldozer, directing the driver. Boards that had once been the BCMA blockhouse were stacked alongside the road, and there was a single strand of barbed wire across the entrance to the slack dump. Hanging on it was a flat square plate with a message from Mr. Fuller, the company, and the steel mill that owned it:

No Trespassing

My blood boiled at the sight of it. "Dad, you gave me this place!"

He gripped the steering wheel and stared at the bulldozer. "You promised to never let another rocket land in Coalwood."

"That was an engineering error, and we've already fixed it."

"Mr. Fuller made this decision," he said, "and he had every right to do it."

"What right? You gave us that lumber, we didn't steal it. The same with the cement for the launchpad."

Dad considered me for a moment, as if trying to decide whether to proceed. "Listen, little man," he said finally, "there's not a damn thing I can do about this. You don't like the way things are run around here? Go to college and then come back here. In a few years, I bet you'd be running the whole place."

My potential for arrogance made a sneak attack on me. I said, "Dad, when I get out of this stinking hole, wild horses couldn't drag me back."

My words were meant to hurt him, and they did. He sucked in a breath and raised his hand at me. I waited, knowing I had gone too far, but no blow fell. He let his hand drop into his lap. "I can't believe you'd say that about Coalwood," he said.

I was instantly sorry. *Damn me and my smart mouth!* Just then, Roy Lee and O'Dell came racing up in Roy Lee's car. They got out and surveyed the scene, looking to me for guidance. I got out of the truck and led them to the stack of our lumber. I picked up a board and they did too. "We're going to build our blockhouse back," I said.

Another car arrived, this one with Sherman and his dad in it. Then another, this one with some of our machinists, and then another with Mr. Dubonnet, Tom Tickle, and some of the other miners. They all stood at the barbed wire. Dad got out of his truck and limped over to where we were. "Homer, this isn't right," Mr. Dubonnet told him.

"This is company business," Dad replied without his usual vehemence.

"We ain't company," Tom said. "We're union."

"You men go on home now," Dad said, but there was no force to his words.

"Not until we build back the boys' blockhouse," Tom said.

The men pulled down the barbed wire and started to cross the slack with the lumber. Mr. Fuller ran up and started yelling at them, but they pushed past him. Mr. Dubonnet stopped at the bulldozer and had a word with the operator, who immediately started to scrape off our pad. I turned at the blockhouse site and watched Dad walk over to Mr. Fuller, who was huffing and puffing and yelling at everybody. Dad touched him on his shoulder and he whirled and went up on tiptoes, yelling in Dad's face. Dad took it for a while and then suddenly reached out, clutched the little man's leather jacket, and lifted him right up off the ground. Roy Lee pointed over at them. "Guess your Dad's negotiating," he chuckled. Mr. Dubonnet looked and laughed aloud.

Mr. Fuller stomped off the slack, and Dad came over and drew me aside. "You've got a wide open field. If you want tubing, machine work, aluminum sheeting, just tell Leon Ferro and I'll sign his requisition. If you fail now, you can't blame me. You'll only have yourself to blame. Understood?"

I grinned at him. "Understood, sir."

22

WE DO THE MATH

Auks XXIII–XXIV

A FEW DAYS later, Mr. Fuller abruptly left town. The fence said Dad had run him out, but I think the real reason was he had done his job for the steel company as the designated hatchet man. A new general superintendent arrived a week later, a Mr. Bundini. Mr. Bundini was a gentleman reminiscent of Mr. Van Dyke, but he also brought with him more bad news from the steel company. The mine was ordered to go to a four-day workweek. Dad met with his foremen and told them a twenty percent cut in salaries—his included—would take effect immediately.

The fall season blew in with a rush of cold wind, the maple trees in the yard flashing orange and then dumping their leaves as if in a hurry to get it over with. It was usually Jim's job to rake up the leaves, but this year it fell to me. It was one of the little things that reminded me my brother was really gone, along with the oddly unsettling silence from his room. His class had scattered. The gossip that reached me said Valentine Carmina and Buck Trant had married. O'Dell said he thought Buck had "lit out for Detroit," gone there, I supposed, to build cars. I worried for Valentine and hoped she was all right.

Jim had stopped taking Dorothy out a few weeks after their first date, a typical move, the chase more important to him than the catch. He had gone off to college with his football scholarship in July. Letters and phone calls from him indicated that he was doing well on the practice field, but needed frequent cash infusions to keep his wardrobe up to college standards. Mom wrote the necessary checks and mailed them off.

Besides my class moving up to senior status, there were other changes at the high school. Big Creek was off football suspension, but Coach Gainer had left for a big school upstate. Big Creek was no longer a powerhouse. The football team lost three out of its first four games.

Dorothy was a majorette in the band, strutting up and down the field in front during our halftime shows. She looked exceptionally fine. Although I hated myself when I did it, I couldn't help but sneak furtive peeks at her every chance I got. She kept trying to catch my eye during band practice or in class, but I refused to give her the satisfaction of looking back. When she cornered me in the hall one morning and started to tell me how sorry she was about Jim, how she just didn't know how I felt, I looked right through her. Then, after she was gone, I looked after her like a lost puppy. I missed her. I admitted it to myself if to no one else, especially her.

With Jim gone, I had access to the Buick on Saturday nights, and Roy Lee and I raced all the way from Coalwood to the Dugout. After we got over the mountain, he never had a chance in his old rattletrap. With two four-barrel carburetors, I could get the Buick up to one hundred miles an hour on Little Daytona. I gloried in the recklessness it took to reach that velocity. It felt so *good*, the big car snorting and the steering wheel shimmying and the bushes and trees on both sides turning into green blurs. After the dance, Roy Lee tried a couple of times to beat me to the drive-in in English. He was better on the curves than me, never hitting his brakes, but as soon as we got on any kind of straightaway, I had him. After a while, he stopped racing with me. "I just wanted to have some fun, Sonny," he said, "but you act like you're trying to prove something." I figured I was just a better driver than him and he couldn't take it.

As Dad promised, all I had to do was to tell Leon Ferro what I needed for the BCMA and it was delivered—steel tubing, sheet aluminum, SAE 1020 bar stock, whatever I wanted. When Mr. Ferro called to let me know of the new materials, he asked for no trade, volunteering to do whatever I asked of him. Everything was in place for us to take the next big step in rocket design. *Auk XXIII* was going to be the first rocket to be based on the sum total of Miss Riley's book, the calculus that Quentin had learned in Mr. Hartsfield's class and that I had learned on my own, coupled with the practical knowledge we had gained through two years of failure and success.

Quentin thumbed over to Coalwood on a Saturday in November, and we went up to my room to work the equations. While Daisy Mae watched us with wide eyes from the pillow on my bed and Chipper from his upside-down perch on the window curtains, Quentin read each procedure aloud from Miss Riley's book, his bony finger running along from equation to equation.

The book described the phenomenon that dictated rocket-nozzle design, and Quentin and I talked about it until we were certain we understood it. When rocket propellant burned, it first produced a river of gas that flowed into the convergent section of the nozzle. If the river continued through the throat at less than sonic speed—that is to say, less than the speed of sound—it became compacted in the divergent section, bound in turmoil, and inefficient. But if the gas river reached the speed of sound at the throat ("The key to nozzle design, Sonny!"), then the gas flow in the divergent section would go supersonic, a very good thing. The series of equations we needed to work described the parameters of thrust coefficient, nozzle-throat area, combustion-chamber cross-sectional areas, and velocity of the gases predicted for any particular propellant.

The book also called for us to make decisions we'd never made before: How high and fast was our rocket to go, and how heavy was our payload going to be? We understood that the questions were related. The first thing Quentin and I did was scratch any payload from consideration. We were committed to the glory of pure altitude. "Let's go for two miles," Quentin said.

"Why not thirty?" I demanded.

Quentin was more cautious. "Let's just see what it takes to double our altitude," he said.

I opened a desk drawer and pulled out a pad of notebook paper. The same equation we'd used to calculate altitude based on time was the one we needed first, good old $S = \frac{1}{2}at^2$.

I did the calculations, assuming our rocket reached maximum velocity immediately upon launch and rounding off the altitude to ten thousand feet. The result equaled a velocity of eight hundred feet per second, or 545.45 miles per hour. When I recalculated, I came up with the same result. This was more than five times faster than the Buick could do on Little Daytona, and I found it difficult to imagine one of our rockets could really go that fast. I shoved the notebook away and threw down my pencil. "This can't be right." I was disgusted. I couldn't even do the first simple equation.

Quentin took a quick look and pushed the notebook back to me. "It is exactly correct. Keep going. Don't lose your nerve."

"I haven't lost my nerve!" I snapped. But I had. The next step was to do the equations for the design of the De Laval nozzle, and privately I quaked at the thought of attempting them. There were dozens of them, intricate, enmeshed, one building on the other—one wrong, all wrong. "You had the calculus class, Quentin. You work them."

"No," he said adamantly. "Miss Riley gave you the book. You know calculus as well as I do. Quit stalling!"

My confidence was gone. Doing those equations was like running a four-minute mile—something possible only for someone far greater than I.

Quentin leaned forward and shook his finger at me. "Listen, old man, if you don't work these equations, what will be the point of all we've done? We might end up building a good rocket that'll fly just fine, and all the grown-ups and teachers will brag on us. Who knows? We might even be able to bluff our way past the judges at the science fair. But you'll know and I'll know—all the boys will know—what could have been done if you hadn't lost your nerve. We could have built a great rocket."

"What's your definition of a great rocket?" I asked.

He crossed his arms and jutted out his chin. "One that does precisely what it's designed to do. It doesn't matter if it only flies two hundred feet. If that's what it's designed to do, and that's what it does, it will be a great rocket." He pointed at the book. "We want our rocket to go to an altitude of precisely two miles. The equations to make that happen are in that book. Do them!"

I looked at the tiny letters and signs in the equations. These were the same equations used by Wernher von Braun and seemed intimate, secretive, his domain. The first equation I needed to do was the one that defined the thrust coefficient. Quentin reached across me and tapped it impatiently. "Are we going to sit here all night?"

"All right, you sonuvabitch, I'll do it," I growled. Quentin sat back and laughed.

The sheets of notebook paper slowly began to fill with my scrawled calculations. For two hours I worked, interrupted only by my mother bringing Quentin and incidentally me some milk and cookies. I got out a straightedge, a protractor, and a compass and carefully drew the nozzle and the casement to the dimensions I had calculated. "Well, here's something," I announced when I was done. I felt achy, my arm muscles and fingers sore from my precise drawings.

Quentin replaced me at my desk. He put his head down and went line by line through my pages of calculations. After an hour of it, he threw the notebook across the room. "You rounded off the powers," he accused. "Your drawings are worthless."

"I forgot how to do it when they're fractions," I said defensively.

"You use logarithms, you twit! How could you forget that?"

Exasperated at my stupidity, I looked up at the ceiling and moaned. "Logarithms!" I was so tired. I just wanted to lay my head down and go to sleep.

"Get back to work!" Quentin snarled.

I could have slugged him, but I sighed instead, got out the differential-equations book with the log tables in them, and went back through all the equations. Daisy Mae got off the bed and crawled up in my lap. She nuzzled my arm and then curled up, occasionally reaching out a paw and touching my chest to let me know she was there. Quentin fell asleep and was soon snoring. After the math, I did the drawings again. Mom had given up calling us to supper. Quentin rose from the bed, stretching and yawning, and again perused my work. Then he carefully stacked the notebook pages, squared them off, patted them, and looked at me out of the tops of his eyes in that significant way he had. "Prodigious work, Sonny."

"You think so?" I said it quietly, but I wanted to shout to the sky my relief.

"I think this is going to be a great rocket."

"Let's show it to Mom," I said. Dad was at the mine and, anyway, I didn't figure he'd care to see what we'd done.

Quentin and I carried the sheath of papers and drawings to her. She was sitting at the kitchen table, sipping coffee, and looking over the new Sears, Roebuck catalog. She put it aside to look at our work. While she did so, Quentin's stomach growled audibly. "What now, boys?" she said after giving each page a careful inspection.

"We're going to build a great rocket, Mrs. Hickam," Quentin told her.

"Before you do that, how about some supper?" she asked him. "Pork chops, brown beans, corn on the cob, and biscuits sound good to you?"

"Yes, ma'am!"

Mom said it was too late for Quentin to hitch home, so he again spent the night with us. I think she just wanted him around. While I watched television, the two of them sat at the kitchen table, talking about this and that. Later, I camped out as always on the couch. Dad came in late and went straight upstairs to bed. I resisted the urge to show him my work.

The following Monday I carried my calculations to Mr. Hartsfield. "For a boy who couldn't do simple algebra," he said after carefully studying the pages, apparently never going to forget my original mathematical sins, "I must tell you I'm impressed. Now, I ask you: What will you do with this? Blow yourself up?"

"No, sir."

He smiled, a facial expression I didn't know he had. "I believe you."

I carried the design to Miss Riley, looking for even more approval. I found her in her classroom at lunch, grading papers. It seemed to me she had come back to school in the fall looking pale and thin. Her eyes, always bright, seemed to be strangely shadowed. Still, she seemed to be having a wonderful time our senior year teaching us physics, using her tiny salary to buy things to demonstrate her lesson of the day: Boyle's Law (a balloon), Archimedes's Principle (flat iron and wooden toy boat), centripetal and centrifugal forces (a yo-yo). The class soaked up everything she had to teach. She looked over my work and praised it. I glowed. "Have you thought about the science fair?"

"We're going for it."

She pulled a tissue from a box on her desk and blew her nose. "Excuse me." She tucked her wool scarf closer to her neck.

"Are you okay, Miss Riley?" I asked, worried for her.

"Just a cold. I always get them this time of year. Come on, let's show Mr. Turner what you've done."

Miss Riley escorted me to the principal's office, and I spread the drawings and equations out on his desk. "An impressive pipe bomb, to be sure," he mused. "I heard you assaulted a softball field in Coalwood a few weeks back. Were there any casualties?"

"No, sir. Well, apart from old Mr. Carson stepping in the hole we had to dig to get it out. He was wandering around in the dark for some reason and sprained his ankle."

"In the queer mass of human destiny, the determining factor has always been luck," Mr. Turner observed, raising one eyebrow to Miss Riley.

"Yes, sir," I said back in confusion.

"The McDowell County Science Fair is in March. Miss Riley believes you should be allowed to represent the school with your . . . devices. The county judges, none of whom believes this school capable of turning out anything but football players, will question you rigorously on your project. They will suspect that you are merely standing in front of a project that your teachers or your parents have actually built. Are you prepared to answer tough questions?"

"Yes, sir."

"All right then. Let's give you an oral exam. What makes a rocket fly?"

"Newton's third law. For every action, there is an equal and opposite reaction."

He stabbed the drawing of the nozzle. "And this peculiar shape? What's it for?"

"That is a De Laval nozzle. It's designed to convert slow-moving, high-pressure gases into a stream of low-pressure, high-velocity gases. If the gases reach sonic velocity at the throat, they will go supersonic in the diverging part of the nozzle, producing maximum thrust."

"You see?" Miss Riley said, grinning.

"You taught him all this, Freida?"

"No, sir. He taught it to himself."

Mr. Turner drummed his fingers on the polished surface of his desk and studied my calculations, slowly turning the pages. "The principal of Welch High School, a tedious man, keeps wanting to make a wager on the science fair. Says he needs the money. It's your decision, Freida. You want this young man to enter, make sure he gets his entry form completed in time."

"Yes, sir!"

As we walked back to her classroom, students thronged the hall, the metallic ringing of lockers opening and closing as they gathered their books for the first class after lunch. Dorothy passed us, walking with Sandy Whitt, the head cheerleader. Sandy gave us a cheerful grin and wave. Dorothy nodded. I very pointedly said hello only to Sandy. After we climbed the steps to the third floor, Miss Riley stopped and leaned tiredly against the wall. "I don't know where my energy is these days," she said, waving away my concern when I reached out to help her. She felt her neck, adjusted her scarf, and then smiled at me with a faint sadness. "By the way, if you see Jake, tell him I said hello and he still owes me that ride to Bluefield."

Jake had been called back to Ohio during the summer. He'd left his telescope on the roof for us boys to use, but I wasn't certain he was coming back to Coalwood. I told her I'd keep an eye out for his Corvette at the Club House, and that seemed to be enough for her. We went back to her room. She seemed grateful to sit down.

WHEN I turned in the drawings based on my scientific calculations to Mr. Ferro, he studied them, asked a few questions, and then called in Mr. Caton to do the work. I rode my bike down to the shop every day after school to inspect how he was doing and sweep up the tailings at the machines, anything to assist. To speed things up, Mr. Caton farmed some of the work out to other machinists. I would have tried my hand on the lathes and shapers, but the machinists were having none of that—this work required too much precision for my clumsy teenage hands.

The black phone often rang for me in the evening, usually Mr. Caton working unpaid overtime on the nozzle. The nozzle was tricky, with its two internal angles that had to meet precisely to form the throat diameter I'd specified. Mom came into the foyer as I bantered with him and shook her head. "The acorn doesn't fall far from the oak, does it, now?"

When the intricate De Laval nozzle was ready, Mr. Caton proudly displayed his work. "You think Wernher von Braun could use me down at Cape Canaveral?"

I thought he could, but I told Mr. Caton I hoped he wasn't going to leave.

"And lose the chance to work for you for nothing?" He laughed, and I noticed for the first time he had a gold tooth.

We chose Thanksgiving weekend for the big test of our new design. Loading the zincoshine was a labor-intensive process, no more than three inches of propellant compressed at a time in the casement, a drying time of four hours required for each segment. With an inner length of forty-five inches, that meant sixty hours of loading across a week. Before I left for school in the morning, I compressed three inches and then another three when I came in from school, and another before I went to bed. The basement smelled like moonshine, and not a little of its vapors spread through the house. "If you come over," Mom told the neighbors, "don't start the rumor I'm running a liquor joint."

I used up almost all of my zinc dust loading the *Auk XXIII*. Getting more was a problem. The BCMA treasury was bare. Still, I wasn't terribly worried about it. I just had this belief that whenever I needed anything to build my rockets, somehow it was going to be there, provided by the Lord or whatever foolish angels had taken on the BCMA as a project. O'Dell said he'd think about a way to get us some money. I said I hoped it would be something better than digging up cast-iron pipes.

On the day before Thanksgiving, Dad solemnly waited for each shift at the portal to call out the names of more men to be cut off. A dozen families were moving out of town, leaving more empty houses to stare vacantly when I rode past them on my bike. Coalwood was unsettled, even spiritually. The church had opened anew, this time firmly in the hands of Methodists from north of the Mason–Dixon line, a dangerous combination, my mother said. The new preacher, a wheedling little man who talked through his nose, twittered from the pulpit about the evils of "corporate greed" and men "who did the devil's bidding." Pooky Suggs, who hadn't been to church in twenty years, said this preacher had it right and gathered a group of men around him to announce a wildcat strike. It lasted one shift, the men creeping to work the next day after Mr. Dubonnet told them to get their butts back down in the mine, but Pooky had gotten a whiff of power and was now muttering dissent on the Big Store steps. "Dubonnet and Hickam are in it together," he declared, passing out his moonshine in fruit jars to the other men. "We got to start lookin' out for ourselves."

Mom roasted a turkey for the holiday, but Dad ate little of it, still obviously upset from having to cut off his men. It was just the three of us around the kitchen table. Jim was home from college, but he had a new girlfriend over in Berwind and had taken the Buick and sped off almost as soon as he deposited his laundry in the basement. He had made first-string on the freshman team, but even that didn't seem to cheer Dad up much. He pretended to watch football on television for a little while after dinner and then put on his coat and walked up to the mine. Mom went out on the porch, staying there sewing and reading magazines with Daisy Mae on her lap, Dandy at her feet, and Chipper on her shoulder until Dad returned, near midnight. I was in my room, designing more nozzles, when I heard her hurry off to bed. I think she didn't want him to know she had waited up for him.

The next weekend, the largest crowd that had ever come to Cape Coalwood showed up, nearly three hundred milling people, even a few from the Welch side of the county. We passed the line of cars that began a quarter of the way to Frog Level with me carrying *Auk XXIII* across my legs in the backseat of Roy Lee's car. It was the biggest, heaviest rocket we'd ever built, four feet in length, and I found myself wishing on the drive down that Mr. Bykovski was around to see it, and Mrs. Bykovski too.

At the Cape, I pressed a cork plug inside the nozzle to hold the Nichrome-wire igniter in place. Some of the new and more curious watchers, not knowing the protocol that had been developed by our steadier customers, came out on the slack to get a better view. I was alarmed to see a few children pull up a log not one hundred feet from the pad, as if they meant to stay there. Sherman went over to shoo them back to the road and then began to round up all the others. Quentin walked downrange to operate the far theodolite, while the rest of us piled inside the blockhouse after running up our flag. We were ready to go.

I was tense as I began the countdown. Although Quentin was confident, I was a little afraid of this big rocket. I took a deep breath and turned the firing switch on the professional-looking console Billy and Sherman had built.

With a mighty burst of fire and smoke, *Auk XXIII* tore off the pad and streaked out of sight, a deep thunder reverberating from mountain to mountain and echoing up the valley. Our audience all looked up after it with their mouths open. So did we boys. If a flock of birds with a sense of humor had flown over at that moment, we might all have suffered. There was no sign of our rocket at all. It had simply vanished. Quentin rang up and reported the same result downrange. A towering funnel of smoke gradually drifted over us. *Auk XXIII* was up there somewhere. What if it came down on the crowd or on us? What if it went uprange and landed in Coalwood again?

"I see it!" Billy yelped. Good old sharp-eyed Billy!

"Where?"

"There!"

It was just a dot, but it grew, and it was downrange, although veering toward Rocket Mountain. It hit the top of a big tree, which shivered from the impact as if to let us know it had caught our rocket. Picking up our shovels, we ran down the slack, the crowd cheering us as we went past.

"Forty-two seconds," Roy Lee cried breathlessly as we ran.

"Seven thousand fifty-six feet!" both Quentin and I called out at about the same moment, both of us capable now of working out the calculation in our heads. It was our highest rocket yet, but it wasn't what my nozzle design had predicted. "What happened?" I worried as we pounded down the slack. "According to the equations, it should have gone three thousand feet higher."

"Don't know," Quentin puffed. "Have to look at the rocket."

Billy led us up the mountain, weaving through the trees and bursting through a line of thick rhododendron into a green glade beneath a ridge. *Auk XXIII* was buried there, up to its fins in soft, wet loam. O'Dell looked around and held up his hand. "Stop, boys," he ordered. "Don't trample this place!"

We pulled up short. "Why?"

He dropped to his knees beside a big oak and dug carefully with his shovel, pulling up a gnarly root. "You know what this is?"

When we all shrugged, he smiled. "Money."

"Not another crazy scheme," Roy Lee groaned.

"No, this one's for real. It's ginseng. This glade's full of it. I've never seen so much!"

"What the hell's ginseng?" Roy Lee asked.

"Indian medicine. People over in Japan and places like that think it cures everything."

"How much is it worth?"

"Well," he said as he dug up another root. "I don't think we're going to have to worry about zinc-dust money for a while."

I had vaguely heard of the stuff being dug around the county, but had never actually seen any of it before. I looked at the dirty ginseng specimen O'Dell handed me, thinking of God and whatever angels He had assigned to the BCMA. "The Lord preserves the simple" was Mom's response when I mentioned this to her.

Quentin and Sherman were busily digging out the rocket, finally pulling it out of the packed earth. Quentin peered inside the nozzle and then ran his fingers inside it, wiping out the greasy residue. "Erosion! The worst we've seen!"

The throat diameter that I had so carefully calculated and Mr. Caton and his buddies had so precisely machined was now an ugly, oblong, pitted abomination. "It ate 1020 bar stock, burned it out like it was cardboard," I marveled.

"We must learn to control this," Quentin said ominously. "Or we might as well quit."

Roy Lee contemplated our sad faces. "Are you two crazy? This rocket just flew almost a mile and a half into the sky. It used to be all our rockets did was lay down and fart."

I poked the tail of the rocket at him. "Just look," I said, my voice tinged with bitterness. "Erosion!"

He reached across the rocket and tapped the side of my head. "There's erosion all right."

THERE were several men in Welch who ran ads in the paper for ginseng root and paid good money, so for once one of O'Dell's schemes paid off. We made enough to buy a full twenty pounds of zinc dust. *Auk XXIV* was ready three weeks later. It was a stretch version of *Auk XXIII,* with a foot of length tacked on to see what difference it would make in altitude. Actually, I had added only six inches of zincoshine. The top half foot was filled with a mix of two-thirds sulfur and one-third zinc, tests demonstrating that it produced a slow-burning oily smoke. We hoped it would help with tracking, but it also meant we'd introduced a half-pound payload. That would reduce altitude.

We tackled the erosion problem too. Mr. Caton inspected the damaged nozzle and suggested a curved throat. It would be more difficult to machine and would require hand polishing, but he was willing to tackle it if I agreed. The theory behind his idea was that the sharp throat I'd designed made for a hot spot along its thin edge. Once melting began, it just kept going, eating out the rest of the throat.

We had our next launch on the same day as the Big Creek Christmas formal. While other boys across the district were washing their cars and going across the mountain to Welch to pick up corsages for their dates, we rocket boys were down on our hands and knees at Cape Coalwood, worrying over the latest *Auk*. Only Roy Lee had a date. The rest of us were going stag. I'd put off asking anyone until it was too late, telling myself it didn't matter because true rocket scientists didn't have time for such things. "How many times do I have to tell you those old boys don't do nothing but chase tail?" Roy Lee said, rolling his eyes. "All those women in bikinis strutting around Cape Canaveral, and there's old Wernher and his boys with their big rockets sticking up in the air. How do you feel when our rockets work?"

"Wonderful!"

"Well, there you are. Rocket scientists feel wonderful too, and who else you gonna want to share wonderful feelings with if it ain't a girl?"

"It is nice to be able to tell somebody when our rockets work," I admitted. I involuntarily thought of Dorothy. It had meant so much to me to tell her about each rocket. Now there was no one except Daisy Mae.

"If you'd got yourself a date for tonight, you could impress the hell out of her about this rocket. She wouldn't be able to get out of her panties fast enough!"

"Roy Lee, you are so full of it."

"Maybe so," he said, grinning, "but I get my share."

There was a cold wind coming down the valley. We ran up our launch flag on the blockhouse and tilted our launch rod accordingly. Billy ran down to the far theodolite this time, and Roy Lee went over to the crowd along the road to chase them off the slack. We were about the only amusement left in town, it seemed. The traffic had increased to the point that Tag had to come down and direct it.

Roy Lee's date for the formal was there, having come across the mountain to be properly impressed. She was one of the majorettes, a shapely, happy girl. I saw Roy Lee put his arm around her. She slapped his hand away when he let it drift down over her breast.

We went through our countdown and I turned the firing switch. There was a puff of smoke at the rocket's base, but nothing else happened. I checked the connections in the blockhouse and tried again. Nothing. I ran a bare wire across the battery terminals and got a rewarding spark. The problem wasn't there. "Don't go out there," Quentin told me even as I walked outside the block-house and studied the rocket with binoculars. "It could be smoldering."

The crowd of people on the road was restless. I saw Pooky come out of the crowd, unlimbering a .22 rifle from his shoulder. He knelt behind a rock and took a bead. "Boss's boy!" he yelled. "You want me to shoot?" With just those few words, I could tell he was drunk.

I waved Pooky off. I thought I knew what had happened. "I think the cork fell out. The curved throat probably doesn't hold it as well."

"So what do we do?" O'Dell worried, looking over at our audience, which was audibly grumbling about the delay. Other men had joined Pooky, and they didn't look entirely friendly. I wondered if they were the unemployed men he led.

Pooky moved in a little closer, holding his little rifle at the ready as if our poor old *Auk* was going to attack him. "Gawddamned Homer's boy. Can't do anything right," he called.

"We've got to go out and fix it," I said.

"And get our butts blown up?" Roy Lee asked. "I don't think so."

"I know what we can do," Sherman said, and he did.

Sherman and I crawled on our bellies, pushing in front of us our makeshift armor of corrugated tin taken from the blockhouse roof. Pooky laughed at us, and the men with him hooted. "Y'all boys look like some kind of silver turtle," Pooky yelled.

I ignored his taunts. When Sherman and I got within arm's reach of our *Auk,* I inspected its base. There was a smudge of soot there, spreading up onto the fins. There was no sign of smoke. The Nichrome wire was lying in the charred remains of the cork, which had either slipped or been pushed out of place by the aborted exhaust. I carefully pulled the wire back and inspected it. It was oxidized and useless. I had carried some old firecracker fuse with me and carefully threaded it up into the nozzle until I could feel it stick in the zincoshine compound. "Sherman, this fuse is three years old. My guess is it's going to burn real fast. As soon as I light it, we've got to make a run for it. Are you ready?"

Sherman nodded. "Ready."

I lit the fuse. It flashed. There was no time to do anything but dig our faces into the slack. Pooky, who had crept in closer, didn't even have time for that. The shock wave of the launch knocked him down, and he got up, howling, and ran to the blockhouse and leapt behind it. *Auk XXIV* zipped out of sight, the crowd *ahhing* and the other boys running out to Sherman and me to pick us up. "Time?" I yelled, my ears still ringing as they peered into our slack-smudged faces.

"Thirty seconds and counting," Roy Lee called.

"Don't see it yet!" Billy reported from his theodolite. "Wait, there it is!"

I looked up at a small, dim, yellow line of smoke. The high-sulfur zincoshine at the top was doing its job, showing us where the rocket was. It was still climbing. "Forty-eight seconds," Roy Lee called when the rocket finally hit downrange, this time on the slack.

"Eighty-five hundred feet," I said, after a moment of mental calculation.

Pooky stood by the blockhouse, brushing his nasty old coveralls off, holding his .22 by the barrel. "You boys are crazy, aintcha?" he said, and spat tobacco juice on the side of our building. He pulled the rifle up and aimed it skyward, jerking off a round. "Now, looky there, I just put somethin' up prob'ly higher'n you." He inspected us, squinting, his yellow upper teeth grimacing in a smile. "Ol' Homer's boy got the money to build rockets while the rest of the town's starvin' to death."

Quentin whooped downrange, and I let myself forget Pooky and raced down the slack. Quentin already had the rocket dug up and was inspecting the nozzle. I stuck my nose in it too. There were pits, but the erosion was much reduced. We grinned at each other. "Superprodigious," Quentin whistled.

THAT night, I sat in the auditorium watching the boys and girls dancing at the Christmas formal. The band, a colored one out of Bradshaw, was lively. All the girls were dressed in pastels, formal dresses with lots of petticoats their mothers had sewn for them or they had sewn for themselves. The boys were in coats and ties. A vision in pink and lace came down the steps from the gym floor and sat down beside me. Melba June Monroe, an eleventh grader, looked me over. She was a pretty girl. I had always liked her. "Hi, Sonny," she cooed. "Boy, is my date boring. I don't even know where he is. Why is a tough little rocket boy going stag to the formal? Do you wanta dance?"

I wanted to dance, and I wanted to take her home in Roy Lee's backseat afterward. Did both, as it turned out. Rocket-boy fame.

IN January 1960, the newspapers ran small articles about the arrival in Charleston of a senator from Massachusetts who was running for president of the United States. His name was John F. Kennedy. Another senator, Hubert H. Humphrey of Minnesota, was also planning appearances in the state. The articles said the West Virginia primary was going to be a battleground between the two men for the presidency. The photographs of Kennedy showed a man with a boyish grin and a pile of hair, and I thought he looked more than a little out of place standing in a crowd of West Virginians, even the polished crowd up in Charleston. When I heard him answer questions on the television shows coming out of Charleston and Huntington, he sounded nasal, with a strange, peculiar accent that wasn't even standard Yankee. I couldn't imagine anyone ever feeling comfortable voting for him. One snowy evening in late February, when even Dad was trapped in the house, he

306 / ROCKET BOYS

suddenly erupted over the newspaper. "Old Joe Kennedy made money bootlegging and now he figures to buy West Virginia for his son. Well, he'll probably be able to. Democrats in this state can be bought about as cheap as they come."

I had sudden insight that maybe Dad and I could talk politics, even if no other subject worked for us. I gave it a try. "If he's rich, maybe he'll bring some money into the state," I offered, speaking of old Joe or John F. Kennedy, take your pick. "We could sure use some around here."

"These people are the worst kind there is in the world," he said. "And their money's dirty."

I thought it pretty good that Dad and I had actually traded a thought. I kept going, saying I thought the worst thing there was in the world had to be the Russians. Didn't we walk around every day waiting for one of their H-bombs to be dropped on us out of the sky, even down here in southern West Virginia? I sat back, waiting for him to tell me how bad he thought the Russians were too, but he surprised me.

"There are Americans I'm a lot more afraid of than the Russians," Dad said. "Like those who think it's okay to use the government to force you to do what's against natural law."

"What's natural law?" I wondered, too late for an answer, because Dad was on a tear.

"Greedy men are the others to keep your eye out for. They'll buy up a company just to run it into the ground. They'll let you work only four days a week, but demand the production of seven."

I knew Dad was speaking from experience. I opened my mouth to reply, but he went on. "Some will tell you that greedy and compassionate men are in competition," he said, "but I'm here to tell you they're not. They run in different but parallel packs, but both will destroy this country before they're done."

I remembered then all his books upstairs in the hall. He had places in his head I couldn't go, and didn't want to either. While I was trying to think up an excuse to leave, Dad leaned forward and predicted, "Dwight David Eisenhower will be the last good president this country will have for a very, very long time."

Just then a bullet pinged through a pane in the living-room window, skipped across the top of his chair where his head had been not a moment before, and lodged in the far wall.

23

SCIENCE FAIRS

Auk XXV

MOM AND DAD looked at the hole in the wall and then at the little crater in the windowpane and talked quietly about what it would take to fix the damage. The bullet had been accompanied by the sound of screeching tires, so apparently whoever had shot at us had taken off down the valley. "A little plaster and paint will take care of it," Dad said of the hole in the wall. "There's glass at the tipple carpentry shop. I'll get McDuff to cut us a pane."

Mom considered the upholstery of the chair. The bullet had left a ragged tear. "I'll scissor off the threads. You won't be able to tell it."

I couldn't believe their calm. I was shaking with both anger and apprehension. Somebody had shot at us! "Let's call Tag!" I demanded. "He'll figure out who did it."

Dad looked at me as if I had just crawled out from under a stump. "If it wasn't Pooky Suggs, it was one of those men who hang around him and drink their courage out of a fruit jar. Tag? We'll just leave him out of this. Wouldn't want to see him get hurt."

Mom silently swept up the glass into a dustpan. At her order, I fetched the vacuum cleaner for her and she ran it underneath the window to make certain she got the tiniest shards, this so the cats wouldn't get them in their paws. Then she put a piece of tape over the hole in the window. Dad went down in the basement and brought back a can of putty and patted it over the hole in the wall. "After it dries, I'll sand and paint it," he said, standing back and observing his work critically. "You won't even notice it's there."

When Mom finished tidying up the living room, she came into the living room and planted herself in front of Dad, who had retreated to another chair and reopened his paper. "Homer, we need to go to Myrtle Beach," she announced. It was a bolt from out of the blue.

Dad looked up from his paper. "We *are* going," he said, frowning. "Miners' vacation."

"I mean before. We can leave this Saturday, stay Sunday, and come back on Monday," she said. "The mine will get along without you for that long."

Dad was still mystified. "Why do we need to go to Myrtle Beach now?"

Mom put her hands on her hips. "I'm going to buy a house down there."

It was one of the few times I had ever seen Dad flustered. His paper deflated into his lap. "Wh-why do we need a house in Myrtle Beach?"

"It's not for us. It's for me. I just need for you to come and sign the papers. They won't sell it to me without your signature."

"Elsie, we couldn't possibly afford to buy a house in Myrtle Beach or anywhere else."

Mom pulled a small black book out of her apron pocket and tossed it in his lap. "That's my savings account over at Welch. For the last twenty years, you've given me your paycheck every month and depended on me to pay all the bills. Well, I've done that, but I've also invested what I could in the stock market. I could buy two houses if I wanted to."

Dad opened the book and looked it over with his good eye. He turned the pages until he got to the last one that had writing in it. "My God, Elsie! Where did you get all this money?"

"I told you. The stock market."

"Stock market? Like in New York?"

Mom nodded. "My broker's in Bluefield. He's been calling me all these years while you're at work."

Dad struggled to make sense of the news. "What stocks did you buy—coal and steel?"

Mom laughed heartily. "Homer, please. Give me some credit. No, I paid attention to what worked with the boys—like Band-Aids. I bought a ton of stock from the company that made the best ones."

Mom had passed Dad's grasp of reality. "Well, I always figured after retirement—"

"I can't wait until then," Mom said sadly. "Anyway, you'll work here until you drop. Not me. I'll be at Myrtle Beach."

Dad was shocked. "You're leaving me?"

"Let's say I'm setting up our retirement house in advance. I'll come back here for holidays, or when you really need me."

"But what about the boys?"

"Jim's gone. I've got plenty of money to pay Sonny's way to college if that's what he wants. I'll stay until the day he leaves."

Dad still couldn't make any sense of it. "But what will the people in Coalwood say?"

My mother's answer startled us both. "Homer," she answered sweetly, "I don't give a shit what they say."

DURING our BCMA meeting in the Big Creek auditorium the next morning, I kept interrupting Billy, sure that my news was more important than whatever it was he was trying to say. I told the boys all about the shooting, concluding, "It broke out the window, went through Dad's chair, and dug a big hole in the far wall." It was only a slight exaggeration. I left out Mom's Myrtle Beach announcement. That still wasn't real to me.

"What kind of bullet was it?" O'Dell asked.

"A twenty-two! Mom dug it out of the wall."

"A twenty-two?" O'Dell laughed. "That's a pop gun."

"Well, it came that close to Dad's head," I said, pushing two fingers together. I was miffed that O'Dell took the shooting so lightheartedly.

Billy finally managed to get his news in edgewise during my continuing elaborations. "Did you hear Miss Riley's sick? Some kind of cancer."

I stopped talking, my mouth hanging open. I felt as if I had been turned inside out. "What are you talking about?" I demanded.

"Mrs. Turner told Emily Sue. Emily Sue told me."

Everything suddenly fell into place. Since Christmas vacation, Miss Riley had stopped standing during class, teaching instead from her chair behind her desk and asking one of us boys to come up and perform her experiments. Once, in February when snow covered the ground, she'd left class and hadn't returned. Mr. Turner came to the classroom to tell us to finish reading the chapter we were on and to be quiet. He gave no explanation for Miss Riley's absence. He looked pale, as if he'd seen something that frightened him, and none of us could imagine *anything* that could scare Mr. Turner.

In physics class that day, I tried not to stare at Miss Riley, but I couldn't help it, and she caught me at it more than once. She looked wan and her eyes were puffy. After class, she made me stay behind to talk to her. "It seemed like you were a million miles away today," she said.

I resisted asking her about her cancer. It had been ingrained in me by my mother to never poke my nose into anybody's business. If Miss Riley wanted me to know, she would tell me. "I was thinking about the science fair," I said. That was at least partly true. It had crossed my mind. I still didn't know exactly what I was going to display or how I was going to display it.

"I've been meaning to talk to you about the fair," she said. "You know, only one name can go on an entry. You put all the boys in your club down on the form. I changed it. You're the one."

"But all of us work on our rockets," I protested. "I guess I'd look pretty bad if it was just me at the fair." I thought about Quentin's hope to be recognized by someone enough to get money for college.

"You'll represent us at the fair," she said firmly. "Because you're the one who knows the most."

"Quentin could do it," I said. "He knows as much as I do."

"Maybe," she smiled. "But you know Quentin. He'd try too hard, probably lose the judges with his vocabulary."

When I just stood there, dumbly watching her, she lost her smile and said, "I presume you've heard I'm sick."

She had dropped it on me like ten tons of coal. "Yes!" I blurted. "What's wrong?"

She began to teach me. "What I have is called Hodgkin's Disease. It's a form of cancer that attacks the lymph nodes. Here." She took my hand and placed it on her neck. "This is where I first noticed that I had swelling. Feel that lump? That's one of the places where it is. I've been tired the whole school year, so I knew something was wrong. The doctors ran tests until they were certain what I had."

I drew my hand away, frightened at touching such an awful thing inside her. I didn't know anything about cancer except it killed you. But then I remembered Dad had survived his colon cancer. Maybe Miss Riley could survive this Hodgkin's thing too. "Are you going to be all right?"

She put an elbow on her desk and cradled her head in her palm. She raised an eyebrow and looked into my eyes. "I don't know. Hodgkin's Disease can go into remission. That means I'll still have it, but I won't be sick. Right now, I'm holding my own."

That sounded hopeful. "Is there an operation?"

"No, nothing to be done. The doctor said to just stay as healthy as I can. Get plenty of sleep, eat whatever I want, that kind of thing. The worst thing is I won't be much help to you on the science fair. After a day of teaching, I just can't keep my eyes open. You're going to have to prepare for it by yourself. Can you do that?"

I didn't know if I could but I said, "Yes, ma'am," anyway. The bell sounded for the next class and she waved me away.

"Keep this to yourself," she said.

I yes-ma'amed her again and hurried down the hall as the last few students slipped inside their classrooms. I went past Mr. Turner's office, the last thing any tardy student wanted to do. He saw me, but said nothing, even nodded at me, his mouth set in a grim line. I guess he knew where I'd been.

"SONNY, I just got told by the union I can't work on your rockets no more," Mr. Caton said on the black phone. "We're out on strike as of tomorrow."

I was stunned. "Did you finish my stuff?" I asked him. He was supposed to be working on *Auk XXV* plus a variety of nozzles, casements, and nose cones for display at the science fair.

"Nope. Ferro put us to work on hurry-up mine projects after he heard about the strike."

"How long's it going to last?"

"About twenty-four hours," Dad said, coming through the foyer. "It doesn't have a leg to stand on."

Mr. Caton overheard Dad. "Your dad's wrong, Sonny. This is no wildcat strike. Union headquarters is behind it. It's going to be a long one."

I told Mr. Caton good-bye and then went to Dad, who had taken his paper to his easy chair—now moved away from the window. "Don't start with me," he warned. "I didn't cause this strike and I can't end it. But the machine shop is open. Caton or anybody else who wants to can work."

"He can't fight the union."

"Well, apparently neither can I."

"Dad, I need your help. I need to be ready for the science fair next week. I've got to have the stuff Mr. Caton was building for me."

He kept his one good eye on the paper, the other one tightly closed. A year ago the town had cheered his bravery. Now, when things were tough, a lot of people said he was just a mean old one-eyed man. "Sorry, little man, but I can't let the union get away with striking over nothing."

I reported the situation to the BCMA the next morning. We talked things over and hatched a plan. It was risky, but it was all I could think to do.

The machine shop was unlocked that night, as it was every night, and Sherman was waiting for me when I arrived on my bike. We opened the front door and switched on the fluorescent lights, which blinked on, bathing the rows of lathes and shapers and drill presses in a harsh, blue-green glare. I found Mr. Caton's lathe and set up a piece of bar stock in it as I had seen him do. I got a cutting tool and put it in the special jig he had fashioned for our nozzles. I turned the lathe on and it whirred up to speed. My first cut was serviceable, but when I tried to make the interior angled cut, the tool wobbled and jammed. "Dammit!" I muttered in frustration. I stopped the lathe and pulled out the bar and threw it on the concrete floor.

Sherman picked up the bar and pondered the wounded piece of steel. "I didn't realize this was so hard."

I wiped the sweat off my forehead with a bandanna. "Neither did I," I admitted.

I changed to a fresh tool and went at it again. I got the first cut, but again, when I started cutting the interior angle, the bar worked its way out of the jig, and the tool dug in and broke. An hour had passed and all I had to show for it were two ruined pieces of steel.

That was when the shop door opened and Mr. Caton crept inside, holding his finger to his lips. I wanted to hug him, but of course, I didn't. He inspected our poor results. "You done pretty good," he whispered. "Yeah, that interior cut's a bugger. Whyn't you boys go on home? Not a word to anybody, okay?"

That was fine by us. Sherman and I sneaked out the door into the cool, damp night. I biked past the Club House and the Big Store and across the bridge and through the silent darkness. As I approached the line of little frame houses on both sides of the creek, I came upon a knot of men gathered on and around Pooky's porch. Others sat on the hoods of trucks parked across the road. "Homer's boy," I heard somebody say, and I put my head down and pedaled faster. I heard a screen door slam behind me, and my heart thumped as I heard footsteps, but then they stopped.

I pedaled on. A car came up behind me and then swerved in front, stopping alongside the road. I recognized the three boys who got out. They were all sons of unemployed miners, including

Pooky's son, Calvin. Calvin was one of the boys who had made a practice of beating me up in grade school every time there was a strike. Well, those days were over as far as I was concerned. There were too many of them to fight, so I pedaled at them and then jumped off my bike and scrambled out of the street light and up the hillside. There was a path there that led to the dirt road that ran to the Coalwood School, and I figured I could hide in the darkness. Once on the road, I dodged into some tall weeds and huddled there until I heard them run past me. "Sonny?" they called. "We just want to talk."

I was too smart to fall for that. After a few minutes of silence, I worked my way back down the path, picked up my bike, and darted behind the houses to another path and made my way home. I crawled in bed, with only a few hours before I'd have to get up and catch the school bus. It took me a while to get to sleep and when I did, Mom and Dad and Miss Riley and Mr. Caton and Calvin Suggs alternately intruded into my dreams. Nothing seemed right, the world askew.

THE newspapers said the strike at Coalwood was just one of many across the county. Since the UMWA was notorious for lacking strike funds, a lot of families were facing a potentially desperate situation. The Salvation Army pitched in as best it could, and commodity food was delivered by the state, but I heard my mother saying across the fence to Mrs. Sharitz that she was afraid people, even in Coalwood, might soon go hungry. The Women's Club made up baskets of food and delivered them. Mom helped to organize the effort, but did not go on the delivery rounds. She knew the people would resent her if she did.

Since it was a political season, and Senators Kennedy and Humphrey were crisscrossing the state running for president, the glare of national scrutiny was settling on West Virginia. A lot of people in the state resented the fact that television reporters were flooding in and sending back reports about how ignorant, poor, and helpless everybody was in West Virginia. Both Senator Kennedy and Senator Humphrey thought they had the solution for West Virginia: active assistance from the federal government in the form of free food, with federal-government jobs to follow. If West Virginia agreed to vote for one or the other of the men, food was apparently going to come into the state in dump-truck loads. When Humphrey was asked what was to become of the unemployed miners, he said they would be retrained, which got a big round of applause from his audience. "Retrained for what?" I wondered, watching television at Roy Lee's house.

"Retrain miners." Roy Lee's mother laughed from the sink where she was washing dishes. "I'd like to see that trick!"

Dad could hardly stand to read the newspaper anymore. I came up from the lab just as he threw down the paper on the kitchen table. "The union will never come off this strike if they think all this help is coming their way."

"Are you ready to go to Myrtle Beach?" Mom asked.

"Next week," he said grumpily.

"That's what you said last week."

"I've got to stick around for the negotiating."

"Nobody's negotiating."

I went upstairs and closed my door and lay down on my bed. Daisy Mae joined me in the crook of my arm. My stomach gave me a twinge and I felt nauseated. I was feeling sick to my stomach a lot lately. Everything seemed to be piling on top of me.

SPRING brought rain, and in 1960, everybody was worried about flooding. The problem was that rainwater collected behind the slack dumps up the hollows. Usually, the company bulldozed the slack dams open, but during a strike that wasn't going to happen. A small slack dam finally let loose in a hollow near Six. I woke up one Saturday morning to see a shallow flood coming down the road past the mine. It continued on down the valley, all the way to the Big Store. It wasn't so deep I couldn't bike through it, but I was soaked by the time I got to Mr. Caton's house. I went around to his back door and knocked, looking around to see if anybody had seen me. Mr. Caton appeared and furtively handed me the nozzles and nose cones in a cloth flour sack. "The casements are in the alley behind the shop, stacked with regular tubing stock," he said.

I nodded. Roy Lee would pick them up the next day and hide them in his backseat.

On my way back home, Calvin Suggs and his two buddies plunged off his porch and came after me, splashing in their bare feet across the road. They almost got me, but I swung the sack over my head and made them duck. It kept them at bay—until the sack slipped out of my hand and went sailing into the creek.

I jumped off my bike and knocked Calvin down with a solid fist to his chest. Astonished, he sat down in the flooded road and watched while I plunged into the swirling, muddy water. The flood nearly swept my feet out from under me, but I kept flailing, frantically reaching down to feel the bottom of the creek. All I felt were rocks, mud, and cold water. When I clambered out empty-handed, I walked right up to him and punched Calvin again. He went down, his nose spraying blood. The other two boys rushed up. I started swinging at them too, and they scattered. When Calvin got up and tried to grab me, I elbowed him in the ribs and he staggered back. "What the hell's wrong with you?" he gasped, wiping his bloody nose with the back of his hand. His left eye was swelling up.

"You made me lose my nozzles and nose cones!"

"Your rocket stuff?"

"Yes, you moron. My rocket stuff!"

The three union boys and I stood in the swirling waters coming down the road and looked at one another. Calvin was definitely going to have a black eye. "Calvin, what the crap are you doin'?" Pooky called from his porch.

"Looks like little Sonny's got your boy's measure, Pook," one of the miners sitting on his porch guffawed.

"Gawddamm you, Calvin, smack 'im, smack 'im good."

Calvin ignored his father. "I'm sorry you lost your rocket stuff," he told me, truly contrite. "I . . . we just wanted to talk to you."

I didn't believe him. I balled up my fists. "Come on, let's get it over with!"

"When the creek goes down, we'll help you look for your stuff," he said, running his hand through his wet hair.

I finally saw that there was no fight in him. I looked at the creek and its swift water. "No need to. It's gone, thanks to you."

"Calvin!"

"Shut up, Paw!" Calvin helped me pick up my bike and then held on to it. "Sonny? When you get out of here and go live at Cape Canaveral, could you maybe help us get jobs?"

The other boys nodded, hope written on their wet faces, their long hair down in their eyes. "It'll be awhile before I'm down there," I said.

Calvin let go of my bike. "That's okay. We'll be here or in the Army. You'll be able to find us."

I rode thoughtfully away, another fragment of the known world gone.

The next day, Mom said she heard a light knock on our front door, and when she opened it she saw Calvin running off. There was a wet, dirty sack on the porch. Inside it were my nozzles and nose cones.

THE next Thursday afternoon, Mom drove me over the mountain to the McDowell County Science Fair in Welch. The Buick was loaded down with the panels, posters, and rocket hardware of my fair exhibit. I had decided to title my entry *A Study of Amateur Rocketry Techniques.* The other boys followed in Roy Lee's car and helped me carry my hardware up to the Welch High School gymnasium, which sat high on the side of a steep hill. Mr. Turner had given us the afternoon off from school. Except for Mom, we were on our own. Miss Riley was out sick.

Nervously, we set up the display, which consisted of a three-hinged fiberboard on which I had taped a number of posters showing nozzle calculations, the parabolic trajectory of our rockets, and the trigonometry we had used to calculate altitudes. I also had drawings of the nozzles and casements and how they worked. Wernher von Braun's autographed photograph sat in a place of honor, and in front, lying down, was the *Auk XXV* casement. Beside it was one of the nozzles that Mr. Caton had surreptitiously built. It was a beautiful construction, the intricate curves glowing silvery in the light.

We made a quick audit of our competition. A Welch High School display of plant fossils found in coal mines seemed our stiffest competition. "Just a bunch of old rocks," O'Dell said. "Nothing to worry about."

I wasn't so certain. Each fossilized plant was identified, and there was also a chart showing the evolution of plants from the dinosaurs to the present day. I thought it was a good job, and I suspected the judges would think so too.

The Pocahontas Industrial Council for Education was the sponsoring organization, a committee created by the businesses in Welch and some of the larger coal mines. O'Dell said the judges were "Welch courthouse politicians," whatever that meant. There were six of them, and they came and stood in a little semicircle around me when it was my turn. They each carried a clipboard. "Which high school are you from, son?"

"Big Creek, sir. Miss Riley's physics class."

One of them squinted at a casement. "You ever blow anything up?"

I thought about Mom's rose-garden fence and mentally crossed my fingers. "We tend to be careful, sir."

"Didn't you set that forest fire over at Davy?"

"No, sir. That was an airplane flare."

"What's that?" another judge asked, pointing at the nozzle, and I got my chance to expound on what it was for, how its dimensions were calculated, and what it did.

A judge peered at von Braun's photo. "I read about you in that grocery-store newspaper. Sounds like you boys do some crazy stuff," he added.

"How high will it fly?" another said, pointing at *Auk XXV*.

"I think around three miles," I said, and then explained how I made that judgment and how we'd measure it when the time came.

The six men rocked on their heels and looked at one another and all *hmmed* at the same time. "Looks to me like this could be really dangerous," said the one who had already called us "crazy." He frowned, wrote something down, and then they went off to look at the other displays.

"Those morons won't let you win," O'Dell griped, coming out from behind the display. "Not after you said you were from Big Creek."

"Looks to me like this could be really dangerous!" Quentin muttered. "Like what isn't, in the pursuit of science?"

At that point, I was just happy to have it over with. I had done the best I could. Mom took us all to lunch at the Flat Iron drug-store. When we returned, the judges were waiting for me in front of my display. The lead judge shook my hand and handed me a blue ribbon. "Congratulations, Mr. Hickam," he said. "Looks like you're going to Bluefield for the area finals."

"I knew we'd win!" O'Dell yelled and took the ribbon out of my hand and, to the amusement of the judges, did a little jig.

Mom stood to one side, a pleased, proud smile on her face. She hugged Quentin when he came over to her.

I was still trying to accept what had happened. I couldn't entirely believe it. We'd won! I couldn't wait to tell Miss Riley—and Dad.

DAD was up at the mine all that evening, inside working with his foreman doing safety inspections. Mom said she'd tell him the first thing the next morning. As soon as I got off the school bus, I headed for Miss Riley's classroom. She was at her desk. When I told her we'd won, she gave me a big happy grin and sent a runner to tell Mr. Turner. The principal tracked me down in history class and ordered me out into the hall. He stared at me. "I just made five dollars off the county superintendent," he said, nearly grinning. "When's the next contest?" I told him the area finals in Bluefield were to take place in two weeks. "Got to get my bet in," he said, and went at a near trot back toward his office.

The win at the County Science Fair made us a little more fa-mous. The Coalwood Women's Club invited us to speak at their monthly meeting in the room above the post office. All our grade-school teachers were there, of course. They beamed with pride.

Quentin and I did most of the talking, boisterously proclaiming how difficult the calculations were and describing how our rockets flew. The ladies breathlessly applauded us. We were next invited to the Kiwanis Club in War and were a hit there as well. The president of the club gave us a speaker's certificate and proclaimed us the "Pride of the Hollows."

Mom and Dad left for Myrtle Beach the following Friday, and I had the house to myself. It was also the weekend of the junior–senior prom. With the Buick gone, Uncle Clarence agreed to let me use his car. I had asked Melba June, the junior that I had danced with at the Christmas formal, to be my date, and she had jumped into my arms, right there in the auditorium. I saw Dorothy sitting alone, watching us. She looked away quickly. Dorothy had a new boyfriend, Roy Lee said, another college guy, but he also said things weren't working out. I made it a special point not to care.

We scheduled a rocket launch on the same day as the prom. Basil's effusive prose summoned our audience:

> *The BCMA will be launching from their Cape Coalwood this Saturday. It is a thing of glory to see, all right. Your reporter has already posted their adventures in this space, but it is worth repeating that just about anything can happen at one of their blast-offs, as witness the one where two of our intrepid boys crept out under the cover of swiftly manufactured armor. . . .*

The people gathered on the road as usual, except this time I noticed they were separated into union and company families, each keeping an icy distance. We ran up our flag and launched. *Auk XXV* peaked out at the predicted fifteen thousand feet, neatly falling downrange, landing with a solid *thunk* near the end of the slack. The impact bent the casement and shattered the wooden nose cone. I had decided to line the interior of the nozzle with a veneer of water putty, an inspiration that I hoped would act as an ablative heat sink. It had worked as I hoped, the only erosion just a few BB-size pits right past the throat. Quentin peered at it and put his hand on my shoulder. "Prodigious, Sonny, prodigious." He looked at me with heightened respect. "You know, every so often I think you really *are* a rocket boy."

An orchid corsage in hand, I picked up Melba June and together we strolled into the Big Creek High School gymnasium, a proud couple. We danced nearly every dance. To my disappointment, Dorothy didn't show. When I took Melba June home, we fogged up the windows good in the Mercury before she gave me a final, adoring kiss and skipped up to her porch, where her parents patiently and secretly waited for her to finish smooching the great rocket scientist and science-fair winner. They opened the front door the moment her dainty little foot touched the first step.

Mom and Dad got back late on Monday after I'd gone to bed. When I came in from school on Tuesday, I found Mom humming around with a contented little smile on her face and Dad in the basement poking into the dark corners of the junk down there, whistling. I had never heard him whistle before, didn't even know that he knew how. "Your dad's quitting," Mom told me when I questioned his behavior. "He's going into real estate at Myrtle Beach. We're moving as soon as you go to college this fall. We're figuring out what we'll take and what we'll leave behind."

Mom must have seen the doubt cross my face, because she rushed to assure me. "He really means it, Sonny. He's had enough. Mr. Butler said he could go in business with him."

That part sounded right. Mr. Butler had been an engineer with the company and then quit to open up a realty business in Myrtle Beach. Dad came bounding up the stairs two at a time, as excited as any time I had ever seen him, with the possible exception of when the West Virginia coaches came to see Jim. "I don't think we need a thing down there," he said of the basement. "We can even leave the washer. Sonny's almost worn out the top of it anyway. We'll buy everything new at the beach." To my astonishment, he hugged her.

I went down to my lab to mix up some more propellant. Daisy Mae made an appearance and climbed up on the counter to watch me work. She started rubbing my arm. I patted her absently, but I was too busy to really pay her any mind. After a while, she gave up, got down, and demanded to be let out. Glad to be rid of the distraction, I opened the door for her.

Later, I went up to my room to do my homework. I was feeling a little sick to my stomach and my head ached too. There was so much to do to get ready for the area fair in Bluefield. I thought about stretching out on the bed, but I remembered Daisy Mae was outside and part of the relaxation was having her curl up beside me. I buckled down to a solid-geometry problem, until I heard a screech of tires that seemed to begin opposite the service station all the way past our house. Whoever it was was in a big hurry. I dug in to the problem again. I heard the storm door slam shut and words between Mom and Dad. "I have to tell Sonny," I heard Mom say, and I knew exactly what had happened.

I came, without being called. Mom waited for me in the foyer, holding my little cat in her arms. Daisy Mae's body was turned awkwardly, her fluffy little feet limp. Her head was lying on Mom's chest and her eyes were half open. There was a trickle of blood leaking from her mouth. I could get no closer. A storm of emotion came out of nowhere and engulfed me. My eyes wouldn't focus, and my mind felt as if it were being sucked down into a whirlpool of red and white swirling blotches. I sat down on the stairs hard and stared ahead. *I let her out* was the first rational thought I had. *I let her out. I killed her.*

Mrs. Sharitz appeared from next door, somehow aware of what had happened. "He's okay, Elsie," she kept saying. "He's okay."

I blinked back to reality, suddenly aware of everything. I got up and ran upstairs and into the bathroom and started to puke. I thought I would never stop.

When at last I felt like I could move again, I came unsteadily down the stairs. The house was deathly quiet. I found Mom alone on the back porch, sitting in a chair. She had Daisy Mae in a shoe box on her lap. How many cats had we buried over the years in shoe-box coffins? Always before, Jim and I had carried them into the mountains to bury them and to give them their last rites, a prayer over a rough wooden cross of birch twigs and twine. For the first time, I missed Jim, missed his strength, his ability to focus on nothing but the task at hand. I went down into the basement and got a shovel and came back to the porch. Mom let me take the box, saying nothing. I set it gently under the apple tree in the backyard and began to dig. Dad came out of the house and stood and watched me and then he left in the Buick, destination unknown. Dandy and Poteet sat nearby, shivering and quiet except for an intermittent whimper. Mom came outside and watched me silently. When I finished burying Daisy Mae, I looked up and all the boys except Quentin were there, called by the invisible network that still seemed to connect everybody in Coalwood.

The boys followed me up to my room and watched me as I sat on the edge of my bed, staring unseeing at the far wall. Roy Lee said, "I'll find out who did this, Sonny, and he'll pay for it, I promise you."

Roy Lee was talking about who had hit Daisy Mae. Until that moment, I thought it had been an accident, but then I realized I had heard those tires squealing once before. Whoever had shot at Dad had murdered Daisy Mae.

I nodded, unable to do more. What did it matter now, anyway? Daisy Mae was gone. I had years yet to live and I missed her already.

I got through my classes during the next week in a haze. Roy Lee drove us to Bluefield for the fair, and together we set up the panels and the displays. Once more, I submitted myself to the judges while the other boys hovered nearby. I gave a little speech and then my answers to their questions, hardly caring whether I won or not. We returned after lunch for the presentations. The third prize and then the second prize were announced. I felt my stomach twinge. Oh, God, *please, no,* I thought. We weren't going to win anything. What an embarrassment to go back to Big Creek with nothing. As soon as we'd gotten outside our little county, we'd been a complete and utter failure. The head judge stood up and leaned on the podium. "First prize goes to—and this is a first for this high school, ladies and gentlemen—Big Creek High School, represented by Homer Hadley Hickam, Jr., for *A Study of Amateur Rocketry Techniques!*"

Quentin couldn't contain himself. He jumped up and yelled,

"Whoo-whoo!" before subsiding in embarrassment. O'Dell danced around with both his hands in the air like a victorious boxer. Roy Lee cackled and then slugged me hard on my arm. Sherman laughed and clapped his hands. Billy sat back in his chair, wiping his forehead in relief. The auditorium burst into applause. I just sat there with a big silly grin on my face. I couldn't believe it. *We'd won! We were going to the National Science Fair!*

"I told you, I told you, I told you," Quentin kept saying to me over and over.

When things settled down, we got yet another award. An Air Force major stood and announced that we had earned a first-prize certificate for being "Outstanding in the Field of Propulsion." He talked about our exhibit, saying it contained the most sophisticated rockets he'd seen this side of Cape Canaveral. "You got that right, Major!" O'Dell brayed.

After the presentations were over, the major came by to shake my hand. He said he hoped I'd consider the Air Force as a career. I had the other boys come over, and when I introduced them, he beamed and said, "The United States Air Force would love to have each and every one of you." He had apparently not noticed Sherman's shriveled leg or Quentin's doubtful slouch.

It was raining as we drove home through the twisting valleys and the coal-smeared towns of Mercer and McDowell Counties, the mines alongside the road so quiet and empty they looked as if they'd been abandoned for a thousand years. A bus passed us. HUBERT HUMPHREY FOR PRESIDENT was emblazoned on its side. It sent back a dirty spray that splashed across the windshield. Roy Lee eased up on the accelerator a bit. A few miles farther down the road, the bus that had passed us with such alarm had stopped. Somebody was out of it, waving his hat at a tiny gathering. I was feeling sick to my stomach again and my head felt like it was splitting. Roy Lee stopped to let me go throw up, and when I came back from the ditch, the other boys had climbed up on a stone wall. Hubert Humphrey was a rotund little man, whose jaw and arms seemed to be connected by a string. The more he waved his arms, the faster his mouth seemed to move. He was on a tear, promising the crowd if he got to be president, the government was going to come in and set everything right, even run things if it had to. No one would ever go hungry if he was president, no sirree, and nobody would lack for a job either.

"Let's ask him a question about space," Sherman suggested, and he waved his hand at him, but Humphrey never looked his way. He was still talking when Roy Lee guided his car past the bus and then sped up, slowing only to make his way through the little town of Keystone, the streets empty except for a mangy dog picking around the front of an abandoned store.

MR. Turner called me to his office the following Monday to shake my hand. "You've surpassed all my expectations, my boy," he said. "I'm going to call an assembly. We're going to give you a proper send-off to Indianapolis."

During the assembly, all the members of the BCMA got to stand up and take a bow. Mr. Turner called me up front. "I'll do my best to represent Big Creek," I said, frowning into the stage lights and trying to ignore my splitting head. I was still having bouts of nausea and now headaches.

Miss Riley stood and said, "This just goes to show that Big Creek students can do just about anything they want to do. I know Sonny is going to make us even more proud in Indianapolis."

Seeing Miss Riley so positive and hopeful made me ashamed and disgusted with my symptoms. I was just being a weak sister.

That night I was drawn to my backyard by some undefinable need. All was quiet there in the dark, save the rustling of the leaves in the apple tree, barely stirred by a gentle wind. I walked deep into the yard beyond the light from the kitchen window and stood very still, scarcely breathing, wondering why I was there, hoping something inside me would give me the answer. The night air was so clear that when I looked up, the stars seemed to form a glowing blue-white bridge that arched from mountain to mountain. I stood enthralled, letting my mind wander happily down the starry trail until my attention was drawn to the fence where I was surprised to see someone standing there looking my way. As dark as it was, I recognized him just by the way he tilted his head. "Roy Lee?" I called. "What are you doing there?"

"I was looking out my window and saw you come out into the yard," he said. "I've been meaning to talk to you, but I wanted it to be kind of private-like, you know? Guess this old backyard is as private as it gets in Coalwood."

I waited while Roy Lee leaned on the fence and had a bout of general fidgeting, clearing his throat, squinting, running his hand through his hair, and so forth. Whatever he had to say, he didn't much want to say it. "Roy Lee, what?" I finally demanded.

"I found out what happened to her." He nodded toward the apple tree and I realized he was talking about Daisy Mae. "I found out who killed her like I promised you I would."

I came over to the fence. "Who was it?" I hissed, ready to commit murder. "Pooky, right?"

"No. But it was one of those blamed idiots who follows him around. Pooky most likely put him up to it."

As mad as I was at Pooky, I surprised myself by thinking first of Calvin. Calvin had been mean to me his whole life but I had come to see him in a different light since he had helped me by finding my rocket stuff. Still, if Pooky had sentenced Daisy Mae to die, I couldn't forgive him, even for Calvin's sake. "I have to do something, Roy Lee," I said. "I can't let him get away with this."

"You don't have to do anything, Sonny," Roy Lee said. "Pooky's left town. The way I heard it, Calvin tried to keep Pooky from beating up his ma and got smacked pretty hard. The neighbors called Tag and Tag came right up, kicked in the door, and threw Pooky in the creek. Tag told him if he ever saw Pooky's face in Coalwood again, next time he'd throw him down the mine shaft. Pooky couldn't get away fast enough, Tag was that mad."

"What about Calvin and his mother?"

"The widow Clowers up in Six took them in until Calvin graduates and goes into the Army."

"And what about the man who ran over Daisy Mae?" I asked.

Roy Lee twitched some more, turning finally to look up at the dark mountains. The lights from the gas station burnished his sleek black hairdo. "I'd tell you his name, Sonny, if you made me, but the way I heard it, he's real sorry now he did it. You want to know?"

I considered it and then shook my head. What good would it do for me to go around sharing the same town with someone I hated? I just didn't see the sense of it. Anyway, I figured the man who killed Daisy Mae would eventually get his due without any help from me. Justice, after all, had finally come to Pooky—Coalwood justice. It seemed as if the town had a way of eventually settling everything if only one was patient enough to let things sort themselves out.

"Thank you for telling me, Roy Lee," I said, and all of a sudden I realized how much he meant to me. I found myself wanting to say that I hoped Roy Lee would always be my friend, and that I could be his, no matter what happened to us or where we went or how far apart we were. I settled for hitting him on the shoulder and then letting him hit me back, a good balled fist to the shoulder that hurt. That said everything I wanted to say without letting the words get in the way of it, anyway.

I said good night to Roy Lee and moved away from the fence and walked to the apple tree, wanting to be near Daisy Mae. I knelt and patted her grave, taking a handful of soil from it. I would put it in a fruit jar and take it to Indianapolis with me. Standing, I took a deep breath of mountain air, and then I knew something else. Mr. Dubonnet had been right that day years ago by the old railroad track when he said I had been born in the mountains and that's where I belonged, no matter what I did or where I went. I didn't understand him then, but now I did. Coalwood, its people, and the mountains were a part of me and I was a part of them and always would be. I also remembered that night when Dad had come back from Cleveland and we had argued in my room. I had gone to my window after he'd left and looked out, envying the men I saw going to and from the mine, because they knew exactly who they were and what they were doing. Standing under the apple tree where Daisy Mae was buried, I realized I didn't have to envy them anymore: I also knew now who I was and what I was going to do. That was when, almost as if someone had pulled a string, my stomach and head stopped hurting.

24

A SUIT FOR
INDIANAPOLIS

"I'LL TAKE CARE of him, Mrs. Hickam," Emily Sue promised my mother from the Buick's passenger seat.

I was behind the wheel of the car, my head down, and I was fuming. Emily Sue and I were going to Welch to buy me a suit for Indianapolis. I didn't see why I needed one. What was wrong with the clothes I planned on wearing, my cotton pants and plaid shirts and penny-loafer shoes? I was supposed to be a young scientist, not some fancy pants like *Peter Gunn* on television or somebody. Besides, my displays and charts for Indianapolis needed work. I didn't have time to traipse over to Welch for clothes.

Emily Sue was already, in her opinion, an adult, unlike certain other members of her class, such as me. It was up to her, therefore, to make certain I would not embarrass Big Creek High School or, for that matter, the entire state of West Virginia in Indianapolis. My clothes, never fancy, were her primary concern. Emily Sue's mother had driven her across the mountain to make her pitch about proper dress at the National Science Fair to Mom, who called me up from the basement, where I was screwing on the hinges of my new display boards. "Take her to Welch," she said, nodding toward Emily Sue, who sat on the couch with a big, pleased smile on her face. "Let her help you pick out a suit."

"What do I need a suit for?" I grumped.

"Because we can't have you at the National Science Fair looking like a hillbilly," Emily Sue said.

Mom lifted her chin. "No, Emily Sue," she said. "There's a better reason."

"What's that?" I demanded.

She laid her eyes on me. "Because I said so."

The Buick swerved back and forth as I steered it through one curve after another. In the seven miles to Welch, there were thirty-seven switchbacks. I hardly noticed them. It was the straightaways that seemed unusual. On about the twelfth turn, I said "Thanks a lot" to Emily Sue as sarcastically as I could.

"Happy to do it," she replied.

At least I had the opportunity to ask Emily Sue about Dorothy. Emily Sue was still very much Dorothy's friend. "Um, so how's everybody in the Honor Society?" It was my way of asking without asking.

Emily Sue was way too quick for me. "Dorothy? She's fine. She misses you and she's sorry you're mad at her, but I don't think she stays awake at night worrying about it. Are you still carrying a torch for her?"

"For Dorothy? Don't make me laugh!"

Emily Sue looked across the bench seat at me. "Did you know you raise your eyebrows when you lie?"

I didn't say anything else to her the whole way to Welch.

It was a Saturday, and Welch was filled with throngs of shoppers. We parked behind the Carter Hotel, paid a quarter to the attendant, and walked down the hill toward Main Street. Emily Sue led me to Philips and Cloony, a men's shop. I hesitated at the front door. "Now what's your problem?" Emily Sue asked.

"I don't want you to go in with me."

"Why? You afraid they'll think I'm your girlfriend?"

"I'm just kind of embarrassed. I can pick a suit out by myself."

She eyed me, a doubtful expression on her face. "Oh, all right," she sighed. "Meet me at the parking building in an hour. And wear your new suit. I want to see it on you."

I agreed, took a deep breath, and went inside. Philips and Cloony was a tiny shop, but it had the reputation of being the best men's store in the county. Its walls were lined with racks of suits and shirts, and it smelled to me like dry-cleaning fluid. When I told the clerk what I wanted, he asked me if I was Jim Hickam's brother. When I said yes, he called the owners down from their apartment upstairs. They were a married couple, a big chunky man and a tiny, bubbly woman. They came padding in with their eyes lit up like cats finding a bunny in the vegetable garden. They said they missed Jim and wondered how they could help me. I told them about the National Science Fair, and they began to lay out brown, blue, and gray suits for me to consider.

They were the kinds of suits the men in Coalwood wore to church. I scratched my head, unsure of myself. Mom had always bought my clothes. Then O'Dell came into the shop. He was in Welch selling more ginseng for our zinc-dust money and had seen me from the street. "Emily Sue's right!" he brayed when I told him my situation. "You need some new duds!"

O'Dell looked through the suits the owners had put out, shaking his head. "Old people's clothes," he said. He went through the racks until he found one in the back he liked. He hauled it out to show me. "Man, you'd look great in this!" he said, and I had to agree. It was the finest-looking suit I'd ever seen.

I tried on O'Dell's suit. It was a perfect fit, and it cost only twenty-five dollars, marked down from twenty-seven fifty. "I'll take it!" I chirped, turning this way and then that while looking in the mirror. The owners looked at each other and shrugged.

I exited the shop, dressed in my fine new suit. I couldn't wait to show it off to Emily Sue. I waved good-bye to O'Dell. He was going to see his ginseng buyer, with two grocery bags full of the root. "After I sell all this, we'll be able to buy enough zinc dust to go to the moon," he promised.

I was a little early to meet Emily Sue, so I walked down to the main street. It was clogged with shoppers. I noticed I was getting some stares from some of them. I guessed they hadn't seen a high-school boy dressed so fine since my brother had left the county. I strutted to the big concrete municipal-parking building, a three-tiered structure that was the pride of the city. It was advertised as the first of its type in the United States, a place where cars were parked on three levels in the same building. I gawked at it every time I saw it. It was too imposing for a Coalwood boy. That's why I chose to park behind the Carter Hotel.

I wormed my way through a crowd of people and saw a table with a JACK KENNEDY FOR PRESIDENT sign on it. Some men were setting up some loudspeakers. Then the martial strains of "Anchors Aweigh" blared out, followed by "High Hopes," sung by Frank Sinatra. "What's going on?" I asked a man putting up a Kennedy poster on a telephone pole.

He looked me over, as if maybe I had two heads, and then said, "The senator's going to make a speech right here in Welch. He'll be here any minute."

Attracted by the music, more people were crowding in. Somehow, Emily Sue found me. She took one look and said, "Oh, my stars!" Her mouth stayed open.

I thought there was something going on behind me that had scared her. I looked over my shoulder, but didn't see anything. "What?" I demanded, turning back.

Her mouth was still open. "What color *is* that?"

"My suit?" I looked at my sleeve. "I dunno. It's sort of an orange, I guess."

"*Orange!* You bought an *orange* suit?"

I shrugged. "Well, yeah . . ."

Just then, a convoy of Lincolns and Cadillacs wheeled into the parking building, their tires shrieking. Emily Sue and I had to step aside or we'd have been run over. We found ourselves right up front in the crowd. "Hey, this is great!" I said.

Emily Sue hadn't even glanced at the signs or the cars. She was still staring at me. "You don't like my suit?" I asked her. "O'Dell came by and helped me pick it out."

She slowly shook her head and then said, "That explains everything."

The crowd was applauding politely as a man got out of one of the Lincolns. He waved and I guessed he was Senator Kennedy. When he was hoisted to the top of a Cadillac, I knew I was right. He was a thin man with a large head and a lot of hair and a brown face. My first thought when I saw him was to wonder how in the world it was possible to get such a tan in the spring. The senator waved again, cleared his throat—somebody handed him up a glass of water, which he sipped—and then he started to talk. The crowd was milling, not everybody paying attention. He was giving it his all though, and I thought it only polite to listen. His speech, delivered with a clenched fist punching out nearly every word, was about Appalachia (which I was surprised to hear we were part of) and the need for the government to help the whole area, maybe, he said, with a TVA-style project. I'd been taught about the Tennessee Valley Authority in high-school history. Mr. Jones said President Roosevelt had used it to help the economy of the hill country of Tennessee and Alabama. I heard my dad say once to my Uncle Ken that the TVA was just socialism, pure and simple. Uncle Ken said it wasn't either, that it was just the government looking out for the little man. Dad had replied that the government didn't look out for anybody but itself.

The senator kept talking. I noticed that his hand crept to his back, pushing in the small of it like it hurt him there. He stood stiffly, like one of Dad's junior engineers after their first day in the mine. His eyes had kind of a sad look to them too. I thought he was in some pain, either in his back or somewhere else.

The Welch audience stayed attentive but quiet as Senator Kennedy promised to create a food-stamp program. The men who had gotten out of the Lincolns and Cadillacs applauded at the proposal, but they were joined by only a few people in the crowd. The senator paused and brushed the hair from his forehead in a nervous gesture. "I think the people of this state need and deserve a helping hand, and I'm going to see that you get it!" he shouted, socking the air. Only silence came back at him. I noticed some people starting to leave. The senator frowned and looked worried, and I felt sorry for him. "How about some questions?" he asked. He sounded a little desperate.

My hand shot up. For some reason, he noticed me right off. "Yes. The boy in the, um, suit."

"Oh, God," Emily Sue groaned. "You're going to embarrass the whole county."

I ignored her. "Yessir. What do you think the United States ought to do in space?"

"Oh, *please,* God," Emily Sue groaned anew.

There was a stirring in the crowd, a few hoots of derision, but Kennedy smiled. "Well, some of my opponents think *I* should go into space," he said. With that, he got himself some appreciative laughter. He looked at me. "But I'll ask you, young man: What do *you* think we ought to do in space?"

As it happened, lately I had been thinking about the moon a lot. In between spring storms, Jake's telescope allowed me to walk down the rills and climb the mountains and stroll the maria of the moon in my mind. It helped me when I was sad about Daisy Mae, or worried about my parents moving to Myrtle Beach, or contemplating my future. The moon had become near and familiar, and that's why my answer just sort of popped out. "We should go to the moon!" I said.

The senator's entourage laughed, but he shushed them with an irritable wave of his hand. "And why do you think we should go to the moon?" he asked me.

I looked around and saw men in their miners' helmets, so I said, "We should go there and find out what it's made of and mine it just like we mine coal here in West Virginia."

There was more laughing, until one of the miners spoke up. "That boy's right! We could mine that old moon good!"

"Hell," another miner shouted out, "West Virginians could mine anything!"

A ripple of good-natured applause went through the crowd. There were a lot of grins. Nobody was leaving.

Kennedy seemed to be energized by the response. "If I'm elected president," he said, "I think maybe we *will* go to the moon." He swept his eyes across the people, now attentive. "I like what this young man says. The important thing is to get the country moving again, to restore vigor and energy to the people and the government. If going to the moon will help us do that, then maybe that's what we should do. My fellow Americans, join with me and we will together take this country forward. . . ."

The crowd responded heartily. Kennedy was talking about working to make the country great again when Emily Sue dragged me away. "What're you doing?" I demanded. "I'm having fun."

"We're going back to Philips and Cloony before they close."

"What for?"

"You're not going to Indianapolis in that orange suit. It's the most carnival thing I've ever seen!"

I stopped dead in my tracks. "I like my suit."

She started to argue, but then said, "I don't doubt it." She put her hand on my back and propelled me forward.

It was after dark by the time we got back to Coalwood. I came inside wearing a dark blue suit, which I hated even as Mom and Emily Sue's mother praised it. Mom said she'd never seen me so handsome. I could only wish she'd seen me in the O'Dell suit. I told her about the senator instead. "You wouldn't believe the things Sonny said to him," Emily Sue sighed.

Dad came in and gave my suit a quick inspection. I told him about the senator. "Kennedy?" He frowned. "A damn pinko if there ever was one."

Dad left, heading outside, probably to go up to the mine. Mom looked after him and murmured, "A good-looking pinko, that's for sure."

After Emily Sue left, I went upstairs and hung my dreary new suit in my closet. At least I had one thing to comfort me. In the jacket pocket was a new tie I'd bought when Emily Sue wasn't looking. It was a glossy light blue, about six inches wide, and painted on it was a big red cardinal, the West Virginia bird. The cardinal was looking up at the sky, and its bright orange beak was open as if it was singing. It was a glorious tie, one that could be spotted clear across a room, which I figured was an important attribute. If I couldn't wear O'Dell's suit, I was still going to show the National Science Fair at least a little Big Creek Missile Agency style.

25

THE NATIONAL
SCIENCE FAIR

THE MONTH FOLLOWING the science fair in Bluefield seemed to flash by. There was so much to do to get ready. Happily, Miss Riley seemed to blossom with the spring. The color returned to her cheeks, and her eyes became bright once more. Every day after classes were done, she worked with me on my presentation skills. She also called teachers in other high schools who had sent students to the nationals, just to get some tips on how to prepare and present. Every day I honed my spiel a little more so that I could quickly deliver a learned presentation on the mathematics of the design of De Laval nozzles, the calculations of specific impulse and mass ratios, and the trigonometry of altitudes needed for an amateur rocketry test range.

Quentin came to my house on weekends and helped me prepare charts and diagrams of the nozzle functions, rocket trajectories, and fin designs. O'Dell found a piece of black velvet somewhere on which to lay out our rocket hardware. He also built some wooden boxes to hold it all, cushioned with newspaper for protection. Sherman and Billy took photographs of Cape Coalwood and put them in a photo album. Roy Lee made three-by-five cards for each nozzle, nose cone, and casement, with a description of its dimensions and function.

While the boys and I kept our eye on the National Science Fair, the Coalwood mine continued to be idle. Some miners, desperate to make some money, tried to enter to work on the hoot-owl shift, but were chased away by union picketers. The company store gave credit until it couldn't anymore. Neither the company nor the union seemed to be in any mood to settle.

Mr. Caton had gone begging to Mr. Dubonnet, and the union chief had unbent just enough to let him build me the nozzles, casements, fins, and nose cones I needed for my display. One set of nozzles showed evolutionary BCMA designs, from the simple countersink version to our latest beauty with an ablative coating. They were all jewels. Mr. Caton had done himself especially proud on one of them, cutting all the waste metal off it until the converging/diverging angles could be seen from the outside. I was certain it looked as good as any nozzle on any rocket taking off from Cape Canaveral.

I went to the union hall and thanked Mr. Dubonnet for letting Mr. Caton do the work. "Just tell them your hardware was built by the UMWA," he said grimly. He had a right to be grim. I knew he was completely out of strike funds and the commodity food from the state was dwindling. I felt nearly ashamed to bother him with my rockets.

At home, Dad still went to work every day, joining his foremen for safety inspections and even rock dusting when necessary. He passed by all our display preparations in the basement going and coming from the mine, but said nothing to me about them. He was gone by the time I got up in the morning, and I was either in my room studying or in bed by the time he got home. He had kept his promise to help me when I asked for it, but seemed to take very little interest in my upcoming trip to Indianapolis. I didn't expect anything else.

On the weekend before I left, I heard Mom pester him about whether he had told the company that he was moving to Myrtle Beach. "I've got to wait until the strike's over, Elsie," he said.

"Why?"

"Because I don't want the union to think they ran me off."

Mom seemed to accept that explanation but I was suspicious. For one thing, since when did Dad care what the union thought? For another, I couldn't imagine Dad leaving without choosing and thoroughly training his replacement. To do that, he'd have to let the company know his plans as early as possible. I also hadn't heard a single peep from the gossip fence about my parents moving. I knew Mom hadn't talked about it, because she considered it a private matter. But it just didn't seem possible that Dad hadn't said something to somebody at the mine about it. Just one little comment would have had the fence buzzing, but all was quiet or one of the boys would have mentioned it to me. So what did Dad really plan on doing? I was too busy to do anything but wonder.

On the night before I was to leave for Indianapolis, Quentin spent the night at my house, not letting me sleep, drilling me incessantly on the details of the trigonometry, calculus, physics, chemistry, and differential equations we used for our rocket designs. Finally, at about three in the morning, I collapsed on my bed and pulled the pillow over my ears. "No more, Q," I begged. "For the love of God, no more."

Through the pillow, I heard him clear his throat. "Is it your plan, then, Sonny my boy, to disgrace the entire state of West Virginia with your confounded ignorance?"

I pulled the pillow away. "Again," I sighed.

"That's the style!" he said brightly. "All right, old chap. An easy one. Define specific impulse."

"Specific impulse is defined as the thrust in pounds of a given propellant divided by its consumption rate."

"And what good is knowing that?"

I let out a long breath. "It's a means of determining the relative merits of propellants. By using the number denoting the specific impulse, calculations can be made to determine the exhaust velocity of a rocket and ultimately its overall performance."

"Good. Now, what do we mean by the weight flow coefficient?"

I let out a groan, stared up at the ceiling, and kept talking. Compared to Quentin, Indianapolis had to be a snap.

THE big Trailways bus at the Welch station accepted my panels and boxes of hardware in its yawning luggage bays. The boys, Mom and Dad, Mr. Caton, Mr. Ferro, Mr. Dubonnet, Melba June, and Mr. Turner were all there to see me off. Basil was there too, scribbling furiously. The *Welch Daily News* had stolen a bit of his thunder with an article about us and our science-fair wins. Basil was determined to outdo the bigger paper by the use of adjectives alone, if necessary.

Emily Sue was also there. She made me open my suitcase and show her my new blue suit. She chucked me on the shoulder like a boy and said, "I guess you'll at least look good."

Just for a moment, I thought I saw Geneva Eggers in the back of the crowd, but when I looked again, I couldn't see her. Dorothy wasn't there either, of course, and I hadn't expected her to be. Melba June gave me a kiss right on my mouth in front of my mother, making me turn nearly purple with embarrassment. "Go get 'em, tiger," she grinned.

Just before I climbed aboard the bus, I was happy to see Jake roll up in his Corvette. I was even happier to see Miss Riley in the passenger seat. Jake got out and clapped me on the shoulder. "Heard you've been doing great things, Sonny. I had to come back to see you off. You know this lady?"

I went over to Miss Riley. She opened the car door and I knelt beside her when she didn't make any move to get out. I wondered if she was feeling tired again. "Show them what West Virginians can do, Sonny," she said, holding her hand out to me.

"Yes, ma'am!" I promised, shaking her hand. We looked at each other, and she pulled me in and gave me a big hug.

Mom patted me on my arm at the door of the bus. "Good boy," she said. Dad shook my hand without comment, scowling because he had just caught sight of Mr. Dubonnet. The two were taking turns giving each other dirty looks as my bus pulled out.

I sank back into my seat and the bus rolled through the night. I slept through most of it and awoke at first light and was startled when I couldn't see any mountains, just a flat plain as far as I could see. I almost felt naked. We arrived in the city around noon, and I unloaded my displays at the cavernous Indiana Exposition Hall. I was directed to an area on the outer edge of the displays with other exhibitors of propulsion projects. I made a quick inspection of their displays and was relieved to find that none of them approached the sophistication of the BCMA's designs. A lantern-jawed boy from Lubbock, Texas, set up beside me. He was wearing a cowboy hat. He had two designs, one of them using plumbing hardware for rocket nozzles, the other a demonstration of an electromagnetic launcher, with little colored lights that ran the length of a track, that sent a little ball bearing flying off at a pretty good clip. We became instant friends. His name was Orville, but he asked me to call him "Tex."

Tex gave me some news, and it wasn't good: "We ain't gonna win nothing up here, Sonny. Look around. All the prizes go to the big, expensive projects."

Feeling small and lost in the huge hall amidst all the hurrying people going to and fro, I walked with Tex through the other displays and saw what he meant. Most of them were huge, complex, and obviously very expensive. One of them even featured two monkeys in a self-contained biosphere complete with oxygen-generating plants and a food-pellet delivery mechanism. I had never seen a live monkey before, and here were two in, of all things, a science-fair display. THE WAY TO MARS, it proclaimed. I was stunned. The boys and girls who built them were, I realized, the competition we West Virginia kids were going to have to face once we went out into the world. All of a sudden, my future seemed cloudy and my shiny new nozzles crude.

"Most of these monster displays are from New York or Massachusetts." Tex shrugged. "Lots of money involved, and these guys are just plain smart anyway. Something else too. The judges don't like rocket projects. They figure them to be too dangerous. I knew when I came up here I didn't have a chance to win anything."

"Then why'd you come?" I blurted out. I could feel the likelihood of ever getting a trophy in the Big Creek display case evaporating.

"Because it's fun. You'll see."

Tex was right. It was fun. He and I wowed the people who came to see our exhibits, telling them about our studies and what it was like to fire off a real rocket. I used my hands a lot and made big, whooshing noises. It was as if I was an actor on a stage, and I found I enjoyed the attention as long as people didn't press in too close. I had that West Virginia need for a certain amount of space between me and a stranger. I noted to Tex that we always had crowds around our displays, more than most of the big, expensive projects. "Sure, we're popular," Tex said, "but that won't impress the judges."

The judges were to make their review on the fourth day of the fair. The night before the third day, we were all treated to a big dinner and then packed off to our hotel rooms. Tex and I had already swapped with our assigned roommates and were sharing a room. We walked the streets of Indianapolis, which seemed to me a huge metropolis with cars whizzing past and crowds on the street—friendly, but too many of them for me to feel comfortable. I also felt a vague discomfort at the space around me, and then I realized I was missing the mountains. In West Virginia, they were always there, setting real, physical boundaries between the towns and the people. In Indianapolis, people from anywhere could just come up and bump into you.

I told Tex what I was feeling and he laughed. "Man, you should come to Texas if you want to know about flat." He told me about life in Texas and I told him more about West Virginia. When I finished, he said I worried him. "You're not up here just to compete in a science fair," he said. "You're up here to win for all those people back in your little town. What are you going to do when you come back empty-handed?" He shook his head. "Man-oh-man. I'm gonna have to think about this one."

The next morning, Tex and I got off the bus to stand in front of our displays for another day of fun. To my astonishment, I found my nozzles, casements, and nose cones gone.

I just couldn't understand it. Nothing in my experience had prepared me for it. How could they be gone? Who could have taken them and why? Tex came over. "You didn't lock up your stuff?"

"I didn't know I was supposed to!" I cried, my voice nearly cracking.

"Where are you from, Sonny? Oh, yeah. West Virginia, I almost forgot." He showed me the wooden case he'd brought with him and the lock on it. "This is a city. You lock up everything." He gave me a sympathetic look. "You need to report this to security. Come on. I'll take you."

When we finally found a guard, he heard me out and then said there had been a bunch of kids who had come in the night before. They had probably swiped my things. I heard what he was saying, but I couldn't believe it. "But why would they do that?" I asked.

The guard looked at me. "Where are you from, son?"

"West Virginia," Tex said as if that explained everything, and I guess it did.

I went back, despairing, to my display. I still had the pictures of all the rocket boys, Miss Riley and the physics class, the machine shop, Mr. Bykovski, Mr. Ferro, Mr. Caton and all the machinists, the mine tipple, my house, the basement lab with Daisy Mae perched on the washing machine, all there along with my pages of nozzle calculations and my autographed photo of von Braun. I still had O'Dell's piece of black velvet and Roy Lee's three-by-five cards. But without the nozzles, casements, and nose cones, my display made no sense. When the judges came tomorrow, I would have nothing to show them. Tex was busy setting up his display. People were starting to come in. I felt paralyzed. Everything that had happened—our rockets, Mr. Bykovski, Cape Coalwood, calculus, even poor Daisy Mae—had all been leading up to this judgment, and now, even though I already knew I wasn't going to win, I had this terrible sense of a chain of inevitable events leading toward some conclusion being broken. "Tex, what am I going to do?" I cried.

Tex stopped working on his display and came over. He took off his cowboy hat and scratched his head. "Reckon that little town of yours has a telephone?"

I had never made a long-distance call in my life. Tex took me to a phone booth and I dialed zero and told the operator the number and yes, this was a collect call. Mom answered and I told her what had happened. She was speechless. "Mom, I've got to get more rocket stuff somehow. Could you talk to Dad or somebody?"

There was a long pause at the other end. "Sonny, the strike's gotten even uglier this week. Some union men chased a foreman off mine property yesterday. Tag's up at the tipple now, guarding it. Your dad's threatening to go punch John Dubonnet in the nose. I heard him tell Clyde the company might call in the state police."

I was desperate. "Mom, I need help."

She sighed. "I'll do what I can."

I felt suddenly foolish and selfish. Here she was telling me the whole town was falling apart, that my dad and Mr. Dubonnet were about to get in a fistfight, and that maybe the state police were going to come in, and I was whining that I wanted my rocket stuff. "Mom," I said, struggling against the part of me that wanted to scream, cry, and beg, "it's okay. Honest. I'm sorry I called."

"No, no, Sonny," she said. "You're right to call. I'll see what I can do. But I'm not promising anything, you understand?"

I hung up and went back to my display. People glanced my way and kept going to the other contestants. I found a box and sat down on it. Anyway, I thought, Tex was right. Nobody in the propulsion area was going to win anything. I would just have to go home and accept the fence-line gossip that I'd been too big for my britches and got what had been coming to me, sort of like my dad all these years.

That night, Tex answered the hotel phone and called me. It was Mom. "Can you get to the Trailways bus station in Indianapolis by eight o'clock in the morning?"

My heart skipped a beat. "I think so."

"There will be a box aboard it for you."

"What happened?"

She laughed, but to me it didn't have a happy ring to it. "Sonny, let it wait for another time."

The next morning, I put on my blue suit and my cardinal tie and had my first taxi ride. After I picked up the wooden crate addressed to me at the bus station, I told the driver I was in a hurry and we went careening through the streets as if we were in the Indianapolis 500. We skidded to a halt in front of the exhibition center, and the driver helped me with the box and we went running to my display area. Tex came over and helped me set up, and I reached in my pocket to pay the driver. He had been looking at my photos and shook his head. "I'm from West Virginia," he said. "You don't owe me a thing except to do good!"

"Got a surprise for you, Sonny," Tex said, his eyes widening a bit at my tie. "I been talking to the committee that runs this thing." He nodded to the other boys and girls in the propulsion-display area and they grinned back at us. "All of us did while you were worrying over your stuff. We told 'em if we didn't get a fair shake, we were going to protest, make up signs and parade around just like students do over in Europe and Japan. Scared 'em so much they agreed to put propulsion in our own little separate category."

I was astonished. "Tex, I hope you win!" I blurted, and then was pleasantly surprised to find I actually felt that way.

Tex looked at my nozzles, nose cones, and casements. "Yours is the class act here, Sonny. Go get 'em." He paused. "Gawd, I love that tie. Where'd you get it?"

Less than an hour later, a dozen adults marched into our area. They were the judges. One of them was a young man who spoke in a Germanic accent. I was flabbergasted when he said he was on von Braun's team. "You mean you actually know Wernher von Braun?" I gasped. I couldn't imagine that. It was like being interviewed by St. Paul or somebody out of the Bible.

He laughed. "I work with him every day." Then he started asking me hard questions. I was ready, my pitch rolling off my tongue. My interpretation of the definitions of specific impulse and mass ratio especially seemed to impress him.

When the other judges were finished with me, the young man turned and said, "You know Dr. von Braun's here today, don't you?"

My mouth dropped open. "No, sir! Where?"

He waved vaguely toward the center of the auditorium. "I saw him last over by the biological-display area."

"Tex, will you watch my stuff?"

Tex laughed. "Sure. Get an autograph for me!"

I took off in search of the great man himself. I wandered the aisles, getting myself lost, asking people if Dr. von Braun was nearby. Always, it seemed, I had just missed him. An hour later, defeated, I returned to my display. Tex regarded me sadly. "Man, I hate to tell you, but he was just here. He picked up that nozzle, Sonny." Tex pointed at the special contoured one Mr. Caton had reproduced. "He said it was a marvelous design and wished he could meet the boy who built it."

I ran in the direction that Tex pointed, but it soon became apparent that Dr. von Braun was gone. Disappointed, I returned to find that I had missed another visit, this one from the judges to leave a certificate of my prize and a beautiful gold and silver medal. Tex pounded my back with the joy of it. He'd come in second, but as far as I was concerned, we'd both won. I went to make my second long-distance call ever.

I STEPPED off the bus in Bluefield to a sea of familiar faces accompanied by applause and cheers, brandishing the surprise medal I had garnered. The first thing I heard was, "The strike's settled!" from Mr. Caton. Before I could ask what had happened or how my hardware got made in time, Roy Lee pulled me aside. "Sonny, Miss Riley's in the hospital."

Mom came over. Dad waited at the Buick. Mr. Dubonnet and Mr. Caton were loading my stuff in its trunk. "Go on with the boys," she said. "I'll tell you everything later."

QUENTIN, Roy Lee, Sherman, O'Dell, Billy, and I crept down the quiet, polished halls of Stevens Clinic in Welch. We found Jake sitting beside Miss. Riley's bed. She was propped up, looking very pale, and there was a tube leading into her arm. "Hi, boys," she whispered to us. "Sonny. Back from the fair. How did you do?"

I showed her the medal. "You did it," she said. "I always knew you would." She found a little smile for each of us. "I'm so proud to be your teacher."

"Miss Riley—" I realized suddenly that I loved her, that I had never known and never would know anyone as good as she.

"Can I hold the medal?" she asked.

"It's yours," I managed to choke out. "We wouldn't have won it without you." I pinned it to her pillow.

She turned her pretty face to look at it. "I just got you the book—"

"You did so much more than that!" I tried without success to swallow the thickness in my throat. I was raging inside. Why had God made her sick? Where was the grace of the Lord that Reverend Lanier and Little Richard talked about? Was this an example of it, knocking down a young woman who wanted only to teach?

When she closed her eyes and seemed to drift away, I looked at Jake. He shook his head and led us outside. "She's just gone to sleep. They keep her pretty well doped up."

"Is she going to die?" I asked, nearly inaudibly. I had trouble saying the words.

He didn't answer me directly. "She'll be back teaching after the doctors build her up a bit. Your medal will give her a good boost, I know."

Jake walked us outside to Roy Lee's car. He split me off from the others. "Don't let this spoil what you did," he said. "You should be proud."

I shook my head. "Jake, it doesn't make any sense."

Jake jammed his hands in his pockets, sighed, and looked up at the mountains. "I'm not a religious man, Sonny. You want parables and proverbs, go to church. But I believe there's a plan for each of us—you, me, Freida too. It doesn't help to get mad about it or want to whip up on God about it. It's just the way it is. You've got to accept it."

"Is that you, Jake?" I asked him scornfully. "You accept things the way they are? Is that why you drink?"

He faced me. "I drink sometimes so I don't have to think," he said. "Other times just because it feels good. There's nothing wrong with that, you know—feeling good. You ought to give it a try sometime, maybe give beating up on yourself a rest."

I sagged inwardly. "You know, I'd give my right arm to be like you, to take pleasure from life."

"I know you love living, Sonny. It shows right through you." He looked around. "These old mountains can weigh anybody down. When you get away from them . . . well, there's a whole other world out there. You'll see."

I thought about what he was saying, about what lay ahead of me. I didn't mean to say it. I thought it and it just popped out. "I'm scared of the future, Jake."

Jake turned toward me, but hesitated. He'd been in West Virginia long enough for our terrible stolidity to rub off on him. Then, with a laugh, he threw his arm over my shoulders and hugged me close. "Old son, we're all scared of *that*."

Gratefully, I leaned against him and thought of Saturday nights long ago when once my father carried me up the stairs.

26

ALL SYSTEMS GO

Auks XXVI–XXXI

June 4, 1960

IT TOOK ME awhile, but I finally managed to piece together what happened in Coalwood after my nozzles got stolen. The boys gave me their version, Mr. Caton told me about his part, and then I heard the rest from my mother. In less than an hour, the fence-line telegraph had alerted the whole town that I was in trouble. Mr. Caton headed for the machine shop, but a line of union men, including Mr. Dubonnet, stopped him. Although Dad said the best thing to do was nothing, Mom made him drive her to the machine shop. Then he saw Mr. Dubonnet and got out of the Buick and went nose to nose with him. Roy Lee said it was as if the two of them were finally where they wanted to be, with nothing to keep them apart.

Mom was ready to let them fight it out, she said, but Mr. Caton pushed in between and said, "Now, look here, you won't find a better union man than me and I know we don't have no contract, but we've got to help that boy. He's not up there just for himself. He's up there for Coalwood."

That was when Mr. Bundini showed up in his jeep and told everybody to break it up. O'Dell said Mr. Bundini had a big smile and went over and tapped Mr. Dubonnet on the shoulder and said, "How about we talk a little, John?"

"Oh, your dad was mad," Mom said, frowning at the memory.

"Martin Bundini just left him standing there and went off with his arm over John's shoulder. He was so mad he started to cough, and that made him even madder."

Roy Lee, who heard most of the inside union talk from his brother, said Mr. Dubonnet told Mr. Bundini there wasn't all that much to talk about, that the men would be back to work as soon as a union–management panel was set up that would approve the list of men who were to be cut off.

"That's not all he wanted," Sherman shrugged, telling me the side he heard from his father, which was the management side. "He also wanted all the men who had been cut off last time to be hired back because it had been done wrong."

"Well, that's all John said he wanted," Mom told me in her version. "But John's smart. He wanted something else, something personal out of your dad. Your dad's smart too. He knew it."

"Mr. Bundini was in his glory," Sherman laughed. "He was being the great peacemaker. But he wasn't telling the union everything he knew either."

"Let's see if I understand you correctly, John," Mom imitated Mr. Bundini's Yankee accent. "The company still says the number of miners who will be cut off. Then this panel agrees who those men will be, based on seniority and union rules. Is that it? And hiring back the men too?"

Roy Lee rolled his eyes. "Your dad finally stopped coughing and grabbed Mr. Bundini and argued with him. 'Don't do it, Martin!' he said right out loud. Mr. Dubonnet was laughing the whole time. He'd already won and he knew it."

I hated that it was my rockets that had made Dad give in to the union. "Don't you worry about that, Sonny," Mom said. "This time he needed to give in."

"Mr. Bundini took your dad aside and they started whispering back and forth," Roy Lee said. "Your dad was shaking his head back and forth just as hard and fast as Mr. Bundini was nodding his up and down."

"It turns out," Mom sighed, "that our Ohio owners had made a big deal with General Motors. It needed coal, and fast. The union could've asked for little pink hearts to be pasted on their lunch buckets and the company would've given it to them. Your dad was caught in the middle."

O'Dell's eyes were wide with excitement. "Then Mr. Dubonnet yells out so everybody can hear, 'Homer signs this time!' "

Mom said, "Oh, your dad got hot! 'You can forget that, John!' he yelled. 'I'm not signing anything!' "

"Mr. Dubonnet had the agreement all ready," Billy told me. "He had this folder under his arm and he opened it and took out a paper and brought it over and shoved it right under your dad's nose."

Mom shook her head. "John told your dad, 'I don't often agree with you, Homer, but by God I trust you. The company will sign anything and then go do whatever it wants to do. But if your name's on it, I know you'll quit if the company tries to pull tricks. You sign it or there's no agreement.' "

"Mr. Bundini signed it right off," Roy Lee said. "Then he told your dad to sign it too."

Mom was up on her ladder, painting in another seagull. At the rate she was going, her sky was going to be filled with them before she was through. "I told your dad to go ahead and sign. What difference would it make, after all? We were leaving for Myrtle Beach, weren't we?"

She put down her brush, climbed down off the ladder, and eyed her work critically. "His look told me all I needed to know. I told him, 'Oh, Homer, I should have known!' "

"Practically everybody in Coalwood was in a circle around the machine shop by then," O'Dell said, his eyes wide with the memory of it. "Some women had even brought card-table chairs and were sitting down to watch. It was like a movie."

" 'Elsie, if I sign this, it's my word. I'll have to stay.' " Mom shook her head, looked out at her rose garden and her telephone-pole-thick fence. "That's what he said. I looked at everybody around us and then at the other boys and then up at these blamed old mountains. Well, what else could I say but what I did? I had to do it for you, didn't I? I said, 'Sign it, Homer.' "

"I'm sorry, Mom," I said. I really was, in a way.

She gave me her look that said she didn't quite believe me.

"Your dad asked me if I was staying with him. I told him if I did, he didn't deserve it. Then you know what he said?"

"No, ma'am."

"He said that was the truth." She poured herself a cup of coffee and then went over and dabbed a little brown paint on a coconut. "Well, how could a woman leave a man who admitted he wasn't good enough for her?"

Roy Lee shrugged. "And that was that. Your dad signed, and then Mr. Caton ran inside the shop and got busy. We boys went in and swept up while he and a couple of the other machinists did the work. People were coming in all the time, hurrying things up. O'Dell built you some new boxes, and I burned rubber all the way across Welch Mountain to make it to the Trailways station in time. When you called to say you won, I swear it was like the whole town cheered. You could hear it all up and down the valley."

I listened to everybody who told me their version of the story and said the same thing to each of them and meant it too. "I wish I'd been there to see it." In all its history, I think it was Coalwood's best moment, even though my dad lost to the union, and my mom was forced to stay a little longer in the hills. Jake had it right. There's a plan. If you're willing to fight it hard enough, you can make it detour for a while, but you're still going to end up wherever God wants you to be.

GRADUATION night finally came, and the Big Creek High School class of 1960 walked proudly down the aisle in the gymnasium to receive our diplomas, the boys in green gowns, the girls in white. Dorothy was our valedictorian. Quentin, his B's in phys. ed. catching up with him, was the salutatorian, tied with Billy. Sherman and O'Dell were in the top ten. Roy Lee and I were back in the pack.

Dorothy made a speech. I stirred uncomfortably when she raised her eyes from her prepared remarks and seemed to be looking directly at me. She said, "I know each of us will always care what happens to every other person in our class. We have been very lucky to have been joined together by a wonderful experience—our three years together here at Big Creek High School. I will never forget . . . you." Then she went back to her speech while I fidgeted.

When Mr. Turner handed me my diploma, he stopped me for a personal word. "You've brought great honor to this school," he said. "Not bad work for a bomb builder."

He had placed my National Science Fair medal in the trophy case of gleaming football awards along with an award certificate that read:

A STUDY OF AMATEUR ROCKETRY TECHNIQUES

HOMER H. HICKAM, JR.
BIG CREEK HIGH SCHOOL
WAR, WEST VIRGINIA

GOLD AND SILVER AWARD
1960

The boys and girls of Big Creek went back to their chairs and held their diplomas and looked at one another, filled with present joy and impending loss. Dorothy left before I could talk to her. I took Melba June to the graduation dance that night. Dorothy wasn't there. I would not see her again for twenty-five years.

AFTER graduation, the BCMA gathered in my room. In a more perfect world, perhaps, everything would have worked out as Quentin hoped and we would have all gotten scholarships because of our win. It didn't happen. Instead, O'Dell, Billy, and Roy Lee took the Air Force recruiter up on his offer. Immediately after graduation, they were headed for Lackland Air Force Base for basic training, and then they would use the GI Bill for college. Sherman said his parents had come up with some money for him to go to West Virginia Tech, and he was going to work for the rest of it. I decided to take my mother up on her offer of an Elsie Hickam scholarship. I was still trying to decide which college to go to, but I thought maybe the engineering program at the Virginia Polytechnic Institute. Quentin may not have gotten his scholarship either, but he said if boys from McDowell County, West Virginia, could win a prize at the National Science Fair, he was sure he could figure out how to go to college even if he didn't have any money. He had decided to enroll at Marshall College in Huntington, West Virginia. He wasn't certain how he was going to pay for it, but he'd figure it out when he got there. Somehow, I knew he'd do fine.

The only thing left for the BCMA to do was to decide what to do with the six rockets I had brought back from Indianapolis. Sherman suggested we split them among us for souvenirs, but Quentin wouldn't hear of it. "Sonny, I've got a great idea," he said. "See, we get this big balloon and fill it full of helium. Then we hang our best rocket from it and let it float up about ten miles and then launch. I've done the calculations. We'll make it into space."

O'Dell had another thought. "Let's make a day of it," he proposed. "Launch from morning to night. We'll put up notices, have Basil write us up, make it a big deal."

"It would be a way of thanking everybody," Roy Lee said.

Sherman and Billy both said they liked it.

Quentin sat down hard on the edge of the bed. "We could have done it, you guys," he said sadly. "We could have gone into space—"

"Aw, Quentin, it's a miracle we ever got anything off the ground at all," Roy Lee laughed. "Let's do this and get out of town while we still can."

FOR the last time, we posted our little notices at the Big Store and the post office. Between the ads for whole chickens and fresh milk in his paper, Basil did us proud:

It is a moment that may well go down in McDowell County history. On June 4, 1960, the Big Creek Missile Agency, fresh from its medal-winning performance at the National Science Fair, is sponsoring a day of rocket launches at its Cape Coalwood range. Everyone reading these words is invited. I tell you this: This writer will be there and with him

everyone I know. There is no more inspiring sight than that of a sleek,
silvery BCMA rocket blasting off from its black, sparkling slack
launchpad, hustling into the sky with a backdrop of green mountains,
splitting the blue sky with its roar as it hurtles high on a great column
of smoke. This may very well be the last chance we will have to see this
grand sight, this amazing sight, this glorious sight. . . .

My basement lab stank one last time with zincoshine preparation, six rockets curing at once. All the boys came to help, and the
fumes of the remnants of John Eye's elixir left us all a little giggly.

I woke early on the first Saturday in June, the day of the final
launches. I moved to my window as I had done so many times
before, to look out at the mountains and the highway that led past
the mine. I half expected to see the usual line of miners making
their way on the path to and from the tipple, and Dad among
them, getting his reports, giving his encouragements and directives,
but the path was empty. The mine hadn't gone to a full seven-day
shift, even with the new orders for coal. I heard the familiar sound
of the backyard gate opening and closing, and there Dad was,
going alone up the path to the mine. He walked hurriedly with his
head down, as if the world depended on him getting to his office
not a moment late. His hands were jammed deep in the pockets of
his loose canvas pants, and his dented hat sat on the back of his
head.

A car came down the road from the Welch direction and turned
right, toward town central. Another followed it, and then another.
When I went to the kitchen and made myself breakfast, I heard the
faint rumble of more cars and trucks passing the house. I thought
for a moment they might be going to our launch, but that wasn't
possible. It wasn't to begin for another two hours. I went back to
my room and put on my summer launch-day clothes—jeans, short-
sleeve shirt, and boots. Before I left, I took a look around my room
and suddenly felt as if I'd returned to it after being gone for fifty
years. There were my shelves, heavy with books and stacks of note
paper filled with the calculations that had defined our rockets.
There was my little dresser and the airplane models on top. Pieces
of rockets, old nose cones, bent casements, and scarred nozzles

were scattered everywhere. The feeling of being gone and then returning was so strong I had to sit down on my bed for a while. In times past, Daisy Mae would have sought me out, rushing to get her head petted and her ears scratched. Nothing stirred. I sat alone, everything quiet except for the sound of cars and trucks passing by.

Roy Lee came to the back door, knocking politely. I met him in the kitchen. Mom was at the kitchen table in front of her beach painting, which she had finally finished. There was a beach house and a woman standing in front of it looking out to sea. "Don't blow yourself up," Mom said, with a look that defied interpretation.

"Yes, ma'am."

Quentin arrived as we were loading rockets in the backseat of Roy Lee's car. *Auk XXXI* was so long we had to roll down the window to fit it in. Quentin and I sat in the backseat, gently cradling the rockets. Billy was waiting with Sherman on the bridge that crossed over the creek to Sherman's house. They wedged into the front of Roy Lee's car. We met O'Dell at the Frog Level crossroads, and he squeezed into the back in between rockets, careful not to bend the fragile fins. We spoke little.

A mile before we got to Cape Coalwood, we came upon the first parked car. Tag was there. He motioned us to him. "Bet there's never been so many cars in Coalwood since it was built. I'm going to park 'em alongside the road, single file. The people can walk in from there."

We were astonished at the number of cars and people. Behind us, more were coming. Roy Lee had put two cases of pop and a gallon jug of water in the trunk to offer our audience. We were going to be a bit short.

Some people saw the rocket sticking out of the window, and shouts of encouragement rang out. "The rocket boys, hoo!" "We're proud of you, boys!" "A-OK, all systems go!"

Some of the people we recognized, but not most. "They're coming from all over the county, looks like," Billy said in wonder.

We drove out on the slack and unloaded our rockets with tender care. Tag seemed everywhere, shooing the curious away from us, turning cars around that violated his single-file parking dictum. I looked up the road past several curves, and the sun sparkled off parked cars as far as I could see. The Coalwood Women's Club was setting up a picnic table with all kinds of pastries and jugs of punch and tea. Tag reserved special places of honor for Sherman's and O'Dell's parents and Roy Lee's mother.

It was noon by the time we were ready to get our first rocket off. We ran up our flag. It was the same one O'Dell's mom had made for us nearly three years before, a little tattered but still serviceable. Wind was negligible. Quentin disappeared downrange trailing telephone wires and carrying his theodolite. We signaled Tag when we were ready, and a hush fell over Cape Coalwood. I looked through the blockhouse portal before beginning the first countdown and saw Miss Riley sitting at the Women's Club table. Two of the Great Six teachers were fanning her. Jake stood nearby, with Mr. Turner.

Auk XXVI had a simple countersink nozzle. It zipped off the pad and flew nicely downrange as if buoyed by the cheers and applause. Three thousand feet, we all agreed, and the altitude was reported to the crowd, which *ooohed* and *ahhhed* appropriately.

Auk XXVII was a one-and-one-quarter-inch-wide, three-and-a-half-foot-long rocket, designed to reach ten thousand feet. When it took off, it jumped from the pad on a silvery column of smoke, stuttered strangely in little puffs, and then seemed to find its way, another spout of fire sending it hurtling skyward. Since it had been the last rocket we'd loaded, maybe the zincoshine had not entirely cured. The crowd, still growing, took no notice of its problems and clapped and yelled exuberantly as it disappeared. It hit with a ground-shaking *thunk* downrange. Nine thousand feet. Not bad for a little rocket not given time to cure.

We hauled out *Auk XXVIII* and set it up. It was designed for fifteen thousand feet. Readying the rockets for launch was hot work, and refreshments were sent over. Mr. and Mrs. Bundini and their beautiful daughters waved at us from the picnic under the trees that shaded the clearing by the road. I saw Mr. Caton and our machinists in a knot. They were working the crowd like politicians, telling their rocket-building stories. Mr. Dubonnet and his union leaders stood nearby, their arms crossed, contented smiles on their faces.

Auk XXVIII worried me for a moment when it bent slightly toward the crowd before straightening up and flying past Rocket Mountain, accelerating on a thick plume of smoke. "It's going to land behind the mountain," Billy predicted, and he was right. We saw it fall, but the noise of the crowd was too great for us to hear the familiar *twang* of steel hitting rock and mountain earth.

I started to tell him to wait, we'd pick it up later, but Billy was off on a run, heading up the mountain. Some of the men from the crowd joined his trek. A half hour later they all came running back with Billy holding the rocket over his head, yellow jackets in close pursuit. The crowd scattered. Jake moved to stand over Miss Riley with a folded newspaper, but the angry bees had too many targets and gave up in confusion, retreating back up the mountain.

Auk XXIX and *Auk XXX* were both designed for twenty thousand feet, but with different dimensions. *Auk XXIX* was two inches in diameter, *XXX* two and a quarter but shorter. *Auk XXIX* was six feet long, the longest rocket we had ever fired. It was such a beauty, I almost regretted having to launch it and see it shattered back on earth. It took off in the mightiest roar ever witnessed at Cape Coalwood, tearing out of a caldron of flame and smoke. Our calculations put it just under four miles. *Auk XXX* vaulted off the pad similarly, its parabola drifting up to twenty-three thousand feet. I looked downrange and saw Quentin out on the slack, joyfully jumping up and down.

Auk XXXI was our last and biggest rocket—six and a half feet long, two and a quarter inches in diameter. We carefully raised it into a vertical position and then lowered it on the launch rod. Inside it was the nozzle touched by Dr. von Braun. It had been designed to reach an altitude of five miles. With a rocket this size, I thought perhaps we were exceeding the critical dimensions of zincoshine. I hoped it wouldn't blow up, but I knew it might. I knelt at its base and started twisting together the ignition-wire connections.

"Sonny," Roy Lee said. "Do you see who's here?"

I looked up from my work. "Who?"

"Look."

Tag opened a path through the crowd, and there stood Dad in

his work clothes. Roy Lee went after him, escorted him out on the slack. I heard Roy Lee say, "Come and help us, Mr. Hickam."

"You don't need my help," Dad said. "I just came to watch."

All the boys protested. "No sir, you can help all you want." "Whatever you want to do, sir, you go ahead and do it."

I stood up, brushing the slack off my jeans. "A rocket won't fly unless somebody lights the fuse," I said. "Come on."

Dad entered the blockhouse, and I directed him to the firing panel after checking the connections. "This one's yours, Dad, if you want it."

There was no mistaking the pure delight I saw spread across his face as he knelt in front of the panel. Roy Lee called from the back door. "Whenever you're ready," he said.

I counted down to zero and Dad turned the switch. *Auk XXXI* erupted, blowing huge chunks of concrete loose from the pad. The crowd took a step backward, and some of them started to run. *Auk XXXI* seemed to split the air that filled the narrow valley, a shock wave rippling across the slack. Women screamed and men clapped their hands to their ears. We boys came pouring outside, Billy at his theodolite, O'Dell with his binoculars. The thunderous din didn't stop. *Auk XXXI* kept pounding us as it climbed. Men, women, and children all watched it with mouths agape, eyes wide, their cheers stuck in their throats.

At the Big Store, those few old men not at the launch got uncertainly to their feet as the thunder reached them. They stumbled into the road, shading their eyes, the trunk of fire and smoke tearing out of the mountains like God's finger stuck suddenly toward the sky. In his church, Little Richard raced to the belfry and began to toll the bell in celebration. Some of the junior engineers down from Ohio were on the Club House roof with girlfriends and Jake's telescope. They raised their beers to what they saw rising from the mountains.

Roy Lee kept his eye on his watch. "Thirty-eight, thirty-nine, forty . . ."

"Still see it," Billy announced, the great spout of smoke turned into a dim yellowish streak. "Just about gone . . ."

"Forty-three, forty-four . . ."

"Gone," Billy announced.

Gone at forty-four seconds. I did a quick calculation. Assuming it was flying nearly vertical, *Auk XXXI* had disappeared at an altitude of thirty-one thousand feet, nearly six miles high. I became aware of movement beside me, and I was astonished to see Dad prancing along the slack, waving his old hat in his hand. He was exulting to the sky. "Beautiful! Beautiful!"

As *Auk XXXI* raced across the sunlit sky on that glorious day, I instead watched my dad, and waited patiently, and with hope, for him to put his arm around my shoulder and tell me, at last, that I had done something good.

"There!" I heard Billy yell. *"There it is!"*

People surged from the road across the slack, following the other boys as they raced after our last, great rocket. Dad stopped his dance and put his hat over his heart. He bent over as if a great weight had suddenly been dropped on his back. He looked at me, his mouth open, and I saw in his eyes a curious mixture of happiness and pain that dissolved into fear. I went to him and put my arm around his shoulder, supporting him while he fought for air. "You did really good, Dad," I told him as a spasm of deep, oily coughs racked his body. "Nobody ever launched a better rocket than you."

EPILOGUE

ALL OF US rocket boys would go to college, something not likely in pre-*Sputnik* West Virginia. Roy Lee became a banker, O'Dell went into insurance and farming. Quentin, Billy, Sherman, and I became engineers. Sherman died unexpectedly of a heart attack when he was only twenty-six years old.

My brother became a successful high-school football coach and a mentor to hundreds of young men, helping them through the difficult transition from adolescence to manhood. Although we had our differences while growing up, I am now, as I have always been, proud to be Jim Hickam's brother.

Dorothy Plunk is a pseudonym but the actual girl I describe in this book went on to become a wonderful wife to a fine gentleman and the proud mother of two daughters, both of whom excelled in the classroom. I would meet the grown-up Dorothy again during a class reunion twenty-five years after our high-school graduation. We danced to "It's All in the Game" that night and, to no surprise to me, I found that I still loved her. Some things never change. Every once in a while, we talk on the telephone. I am her friend.

As we hoped and prayed, Miss Riley's disease went into remission. When it returned several years later, she continued to teach even when it was necessary for her students to carry her up the steps to her classroom. Freida Joy Riley died, barely thirty-two years old, in 1969.

John Kennedy had two great visions in his presidency: one to go to the moon, the other to fight for freedom across the world. I believed equally in both, so I volunteered for Vietnam, delaying my dream of working on spaceflight. The irony was not lost on me when I climbed out of a bunker one morning and found a dud Russian 122-mm rocket buried nearby. I inspected its nozzle and thought it crudely designed.

I never got to meet Wernher von Braun. After building the rocket that took his beloved adopted country to the moon, he died of colon cancer in 1977. Vietnam and other work delayed me, but in 1981, twenty-one years after the BCMA fired its last rocket, I finally grasped the dream of my youth and became a NASA engineer at the Marshall Space Flight Center in Huntsville, Alabama, Dr. von Braun's old headquarters. Over the years that followed, many of the men and women on his team became my colleagues and friends. I trained astronauts, talked them through their science experiments while they were in orbit, and often went to Cape Canaveral for launches of the space shuttle and other rockets. I went to Russia and sat across the table from the men who launched *Sputnik I,* and worked with men and women from Japan, Canada, Europe, and across the planet who shared the vision of space exploration with me. My career with NASA was everything I hoped and dreamed it would be.

My father resisted his black lung and kept going into the mine. When the day came that I inherited his books, they included a few of poetry, which surprised me a little. Some of them even had coal dirt on them, enough so I knew he had taken them inside with him. While everybody who knew him figured he was in the mine studying the roof or worrying over ventilation at the face, I wonder now if he wasn't sitting alone in the gob on an old timber with a book of poetry illuminated by his miner's lamp. Which poems he enjoyed there I am not certain, but of all of them that were blackened by coal, he circled but one:

> *have you ever sat by the railroad track*
> *and watched the emptys cuming back?*
> *lumbering along with a groan and a*
> *whine—*

smoke strung out in a long gray line
belched from the panting injun's stack
—just emptys cuming back.

i have—and to me the emptys seem
like dreams i sometimes dream—
of a girl—or munney—or maybe fame—
my dreams have all returned the same,
swinging along the homebound track
—just emptys cuming back.
—Angelo De Ponciano

After his forced retirement at age sixty-five, Dad stayed as a consultant with the company for five more years, living in the Club House. My mother moved to Myrtle Beach, finding at last the scene in her painting. When his lungs finally gave out and his miners refused to let him on the man-trip anymore, Dad joined her there.

In 1989, after completing a tiring mission, I took an extended vacation to the Caribbean. Before I left, I called my mother and father. Mom got on the phone and said Dad's black lung had worsened. He was also depressed, Mom said, because the Coalwood mine had been shut down, the fans turned off, and the pumps deactivated. The mine was filling with water and would never be reopened again. When Dad talked to me, his voice was weak, but otherwise he sounded the same—confident and needing nothing, certainly nothing within my power to give him. Mom got back on the phone and said that, by all means, I should go on vacation, that everything was going to be fine, not to worry. I left them all the necessary telephone numbers where I could be reached and went ahead with my plans. While I was gone, Dad died. When I got home, Mom had already cremated him and scattered his ashes on the ocean she loved. I flew to Myrtle Beach and found her as I had always known her: fully in control and careful that I not be inconvenienced by my father, even in death.

I felt an odd serenity that Dad had died without my even being aware of it. In the nearly thirty years since I had left Coalwood, he and I had not been close. On trips back to Coalwood and later to Myrtle Beach, we gave each other warm greetings, spoke of the weather or the time to drive from my home to his, and left it at that. It was the way he wanted it, and I complied. Such visits, in any case, were for the purposes of visiting my mother, to present myself to her as the years went by for her inspection and approval.

I visited the hospital where Dad had spent his last hours. A sympathetic orderly talked to me, understanding that I wanted to be spared nothing. I had a need to know, even if the purpose of such knowledge was not evident to either of us. The medical answer was what I expected to hear: Dad had suffocated, the macroscopic coal and rock dust that clogged his lungs finally denying him even a scintilla of air.

The orderly spoke of my father as a little man, but he was not, not until his black lung made its final assault. In a space of a few short weeks, he had shrunk, literally collapsing around his lungs as they became the entire focus of his being.

And he had struggled. The orderly, and another, had to hold him down on the emergency-room gurney so the doctor could do what little there was to be done. Dad had clawed at his throat and his chest as if to rip them open. The orderly said his eyes stayed open to the end, and I could visualize those steel-blue eyes blazing. He had been alert until he died, the orderly said, and had shaken his head at the last, as if refusing a helping hand. I hoped he had fallen away with his mind intact and felt the warm envelopment of darkness, as if perhaps he had returned to his beloved deep mine for one last time. I hoped that the hand that had reached for him had been perhaps one of his foremen trying to bring him from the darkness into the light, and that he had recognized him and had at last reached out, had taken what was being offered.

But I doubted it. It wasn't his way.

I made certain Mom was settled back into the house she loved, near the great brown strand of beach that runs along the Atlantic. She and Dad had recently gone through all the things they had carted off from Coalwood and boxed up items they felt either my brother or I might want. I came home, stuffed away the boxes she had given me, and my life kept going as if there had been no

change. I had my work to do. The months changed to years. As
time went by, I found myself thinking of Dad more and more, and
I was troubled. Why had his death not caused me more pain? Why,
instead, did I feel this odd sense of completion and reconciliation,
as if everything between us had been settled long ago?

Feeling a need to connect with a past I had all but forgotten, I
started to open the boxes Mom sent with me. All of them but one
had Mom's handwriting on it. I recognized Dad's crabbed scrawl
on a small brown cardboard box. *Sonny,* was all he had written. I
opened it and there I found, carefully folded within layers of tissue
paper, what I thought had been lost long ago—faded ribbons and
medals and a strange artifact, a perfectly crafted steel De Laval
rocket nozzle.

In November 1997, just before I retired from NASA, Dr. Takao
Doi, an astronaut friend of mine, carried aboard space shuttle *Co-
lumbia* one of my science-fair medals and a piece of the *Auk* nozzle
my father kept for me. It was a perfect launch, and as I watched
the great ship blast off from the Cape Canaveral pad, I was filled
with pride and happiness: The BCMA was finally going into space.

Sometimes now, I wake at night, thinking I have heard the
sound of my father's footsteps on the stairs, or the shuffling boots
and low murmur of the hoot-owl shift going to work. In that half-
world between sleep and wakefulness, I can almost hear the ringing
of a hammer on steel and the dry hiss of the arc welder at the little
machine shop by the tipple. But it is only a trick of my imagination;
nearly everything that I knew in Coalwood is gone. Many of the
miners' houses there are deserted or falling into decay. The Club
House sits in tired dilapidation, its roof no longer safe for
telescopes. The great slack dump we called Cape Coalwood has
been bulldozed, overgrown, browsed now by deer, silent to boy-
hood voices. The deep mine is abandoned, its tunnels flooded, the
equipment inside covered with black water. Nothing commemo-
rates the site, only rubble and faded signs in overgrown thickets
where hundreds of men once toiled and sometimes died.
Coalwood's industrial symphony is forever stilled. All that remains
are distant echoes and husks of what used to be.

Yet I believe for those of us who keep it in our hearts,

Coalwood still lives. The miners still trudge up the old path to the tipple, and the people bustle in and out of the Big Store and gather on the church steps after Sunday services. The fences still buzz with news and gossip, and the mountains and hollows echo with the joyful clamor of childhood adventures. The halls and class-rooms of the old schools still hum with the excitement of youth, and the football fields yet roar with celebration on cold fall Friday nights. Even now, Coalwood endures, and no one, nor careless industry or overzealous government, can ever completely destroy it—not while we who once lived there may recall our life among its places, or especially remember rockets that once leapt into the air, propelled not by physics but by the vibrant love of an honorable people, and the instruction of a dear teacher, and the dreams of boys.